ASTRONOMY

FOR BEGINNERS™

JEFF BECAN

ILLUSTRATED BY SARAH BECAN

Writers and Readers

WRITERS AND READERS PUBLISHING, INC.

c/o Benay Enterprises, Inc.
62 E. Starrs Plain Road
Danbury, CT 06810
1-800-860-2139
www.writersandreaders.com

•

Text Copyright: © 2004 Jeff Becan
Illustrations: © 2004 Sarah Becan
Cover & Book Design: Terrie Dunkelberger
Book Editing: Merrilee Warholak and Dawn Reshen-Doty

A Writers and Readers Documentary Comic Book
Copyright © 2004
ISBN #0-86316-999-6 Trade
1 2 3 4 5 6 7 8 9 0

Manufactured in the United States of America

Writers and Readers—
publishing FOR BEGINNERS™ books
continuously since 1975:

1975: Cuba • 1976: Marx • 1977: Lenin • 1978: Nuclear Power • 1979: Einstein • Freud • 1980: Mao • Trotsky • 1981: Capitalism • 1982: Darwin • Economists • French Revolution • Marx's Kapital • Food • Ecology • 1983: DNA • Ireland • 1984: London • Peace • Medicine • Orwell • Reagan • Nicaragua • Black History • 1985: Marx Diary • 1986: Zen • Psychiatry • Reich • Socialism • Computers • Brecht • Elvis • 1988: Architecture • Sex • JFK • Virginia Woolf • 1990: Nietzsche • Plato • Judaism • 1991: WW II • Erotica • African History • 1992: Philosophy • Rainforests • Malcolm X • Miles Davis • Islam • Pan Africanism • 1993: Psychiatry • Black Women • Arabs & Israel • Freud • 1994: Babies • Foucault • Heidegger • Hemingway • Classical Music • 1995: Jazz • Jewish Holocaust • Health Care • Domestic Violence • Sartre • United Nations • Black Holocaust • Black Panthers • Martial Arts • History of Clowns • 1996: Opera • Biology • Saussure • UNICEF • Kierkegaard • Addiction & Recovery • I Ching • Buddha • Derrida • Chomsky • McLuhan • Jung • 1997: Che • Eastern Europe • Lacan • Shakespeare • Structuralism • 1998: Fanon • Adler • Gestalt • History of Cinema • U.S. Constitution • English Language • Post Modernism • 1999: Stanislavski • Body • Castaneda • Krishnamurti • Sai Baba • Scotland • Wales • 2000: Artaud • Bukowski • Art • Piaget • Eastern Philosophy • 2001: Dante • Garcia Lorca • Garcia Marquez • Iris Murdoch • Rudolf Steiner • 2002: Marilyn • 2003: Astronomy

---

ASTRONOMY
FOR BEGINNERS™

CONTENTS

INTRODUCTION

Astronomy,

the study of our universe, could very well be the oldest science in, well - the universe! On one level we can say that it began the moment early humans started observing the Sun, the Moon, the stars and the planets, and the patterns these objects make in the heavens.

ASTRONOMY HAS THEREFORE ALWAYS BEEN AN **OBSERVATIONAL** SCIENCE.

But, while science was built on observation, it has since come to include much more. After all, observation alone can sometimes trick us.

REMEMBER THAT, TO EARLY HUMANS, IT WAS QUITE OBVIOUS, THROUGH OBSERVATION, THAT THE EARTH WAS FLAT.

But today's scientific methods have "proven" otherwise.

Nevertheless, astronomy proceeded on the level of pure observation for thousands upon thousands of years before people finally crossed the threshold from prehistory into history.

Anthropologists and archaeologists have made highly educated conclusions about what went on before the invention of writing.

However, when the Sumerian civilization of Mesopotamia invented writing sometime around 3500 BC, they invented history as well. And when they began to record the events and facts they considered important, they included their observations of astronomy, which, by that time, were already quite sophisticated.

MeSoPoTAMiA
(IRAQ)
TiGRiS
(IRAN)
EUPHRATeS
(EGYPT)

AREN'T YOU GOING TO WRITE DOWN HOW FAR I CARRIED YOUR SORRY #@! YESTERDAY!!!?

3

As profoundly important as their achievements were, the Mesopotamians' only explanations for their astronomical observations were in the realm of **astrology.**

That is, rather than suggesting natural explanations for the observations, they instead believed there were supernatural explanations for how the patterns of the heavens explained and predicted human events on Earth.

The Mesopotamians led the world in astronomy and astrology for thousands of years. However, when the ancient Greeks became intrigued - sometime around 500 BC - they added an essential new element to observation:

THEORY.

The Greeks constructed conclusions, explanations and predictions based on natural phenomena and *about* natural phenomena,

thereby sending science on a

The ancient Greeks had different theories for everything, which were endlessly tossed about as something of a national past-time!

giant leap forward.

Some people thought the Earth orbited the Sun;

Some people thought that everything remained the same.

Some people thought that everything in nature constantly changed;

Most people thought the Sun orbited the Earth.

(You get the idea!)

Arguments and debates raged on and on.

The Greek tradition of debating theories led the world in just about all areas of thought, including astronomy. In time, their intellectual traditions were passed on to Rome.

After the Roman Empire collapsed around 410 AD, taking a great deal of knowledge with it, the Dark Ages followed, in which many intellectual traditions were lost.

That didn't begin to change until about the 14th century.

Fortunately, the traditions and the spirit of the classical world began to return in the 14th century with the Renaissance (from the French word meaning 'rebirth'). At this time, science finally reemerged, and continued its bold march forward. Moreover, the thinkers of the later Renaissance emphasized yet another important element beyond observation and theory: **experimentation.**

From this time forward, people used **systematic** means to test their theories.

THUS THE SCIENTIFIC METHOD WAS BORN.

Put very simply: The patterns and phenomena of nature are observed. Theories are then made to account for these observations.

Next, these theories are tested.

If a theory fails a test it is disproved and discarded.

If a theory passes a test, then it must pass another.

If it continues to pass more and more tests, the scientists consider it increasingly likely that the theory is true.

If a theory passes all the tests and is also backed up by more and more observation then, at some point, we can comfortably say that the theory is **confirmed.**

The material in *Astronomy for Beginners* represents our present-day knowledge of the universe, as revealed by the sciences of astronomy, physics, mathematics, chemistry, biology, geology and others. So far, the explanations presented here have passed all the tests. A new discovery, a new insight, a new experiment could, of course, always come along and compel us to modify our theories. But at this point in time, most experts in the field are quite confident in our current level of understanding - and also quite confident that there is always much more to learn.

Disregarding the arrows in the figures above, which horizontal line is longer? Our initial observation might lead us to think that Line A is longer than Line B. But a simple experiment to test this theory - by measuring the lines with a ruler - reveals that both lines are equally long. Observation, theory, and experimentation:

science in its purest form!

So the scientific method is the primary intellectual tool of modern science. But what about the more specific, technological tools of modern astronomy?

THIS IS A BOOK ABOUT **ASTRONOMY**, RIGHT?

Well, from prehistory until the start of the 17th century, the primary tools used by most astronomers were their eyes and their brains.

However, in 1609 the first telescope was invented. The telescope magnified the light from distant stars and planets, so astronomers could observe the heavens more closely. We've since learned that the light from distant objects in space can tell us quite a lot about those objects.

We've also learned that there's much more to light than meets the eye.

when ordinary sunlight passes through a *prism* - a glass object with a triangular shape - it emerges in the colors of the rainbow. This had been observed for some time, but it was the English physicist **Sir Isaac Newton** (1642 - 1727) who first came to understand why.

ISAAC NEWTON

HIS THEORY: Rays of white light are actually made up of separate, colored rays of light, which our eyes normally see together as white. The prism bends, or *refracts*, the light and physically separates it into its component colors: red, orange, yellow, green, blue, indigo, and violet.

PRISM

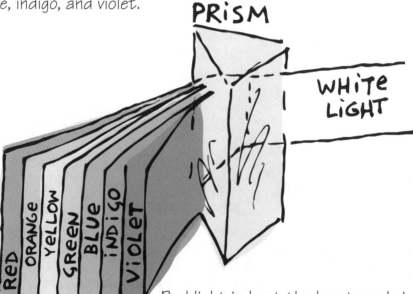

WHITE LIGHT

RED ORANGE YELLOW GREEN BLUE INDIGO VIOLET

SPECTRUM

Red light is bent the least, and violet light is bent the most. Moreover, a rainbow itself is also the refraction of white light, bent by raindrops or water vapor into its component colors.

Experimentation

— which, in this case, is just a fancy word for playing around with prisms! Newton found that if he isolated any single ray of colored light and passed it through another prism, its color wouldn't change. That is, it could not be refracted, or reduced, any further. Newton also found that, if he were to send all of the refracted rays of colored light together into another prism, they would then recombine and reemerge as white light. The more Newton experimented, the more his theories were confirmed. What our eyes perceive as regular white light is actually a collection of all of the primary colors.

Once again, appearances can be deceptive.

11

Stars like the Sun are massive objects. They have so much mass compressed so tightly together that they generate staggeringly high temperatures, which in turn generate constant nuclear reactions. These nuclear reactions release immense

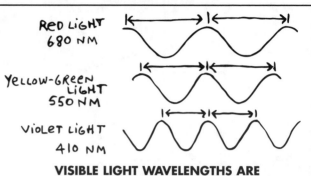

RED LIGHT
680 NM

YELLOW-GREEN LIGHT
550 NM

VIOLET LIGHT
410 NM

VISIBLE LIGHT WAVELENGTHS ARE MEASURED IN THE DISTANCE FROM THE TOP OF ONE WAVE TO THE TOP OF THE NEXT WAVE. "NM" STANDS FOR NANOMETER, WHICH IS ONE BILLIONTH OF A METER.

amounts of energy in the form of **photons** - subatomic particles of light. These particles travel in waves, and these waves come in different sizes, or different *wavelengths*. White light, as we've seen, can be separated into a spectrum of colors, and each color has a different wavelength. Red light has the longest wavelength, and violet light has the shortest.

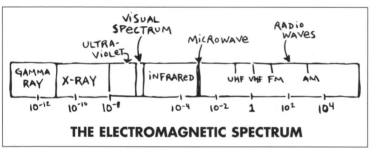

VISUAL SPECTRUM

ULTRA-VIOLET

MICROWAVE

RADIO WAVES

GAMMA RAY | X-RAY | INFRARED | UHF VHF FM | AM

10^{-12} 10^{-10} 10^{-8} 10^{-4} 10^{-2} 1 10^{2} 10^{4}

THE ELECTROMAGNETIC SPECTRUM

But here's where things get even more interesting. The visible light that we can see with our eyes is only one tiny fraction of all of the light - all of the energy - emitted by the Sun. There are wavelengths of light that are shorter than visible light: we call them gamma rays, X-rays and ultraviolet light. There are also wavelengths of light that are longer than visible light: those are called infrared, microwaves and radio waves. All of these different wavelengths of light together make up what is known as the **electromagnetic spectrum**. And because these different wavelengths have different properties, detecting them can give us different kinds of information about any astronomical object that emits or absorbs them.

The standard optical telescope, one means of detection, magnifies a star's visible light. Most telescopes use lenses but some, called *charged-coupled devices*, are computerized and digital. A *spectroscope* is a telescope that can break up the spectrum of visible light

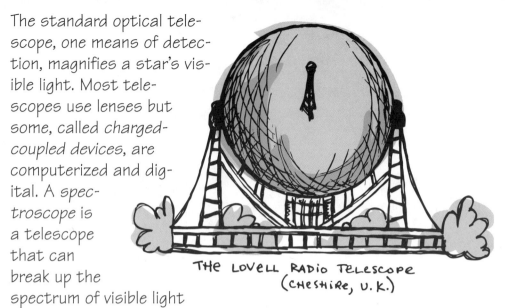

THE LOVELL RADIO TELESCOPE
(CHESHIRE, U.K.)

into its component parts. By analyzing the variations in the colored bands of light emitted or absorbed by an object, we can discover that object's temperature and chemical composition.

Furthermore, *radio telescopes*, which look like giant satellite dishes, are telescopes designed to receive an object's radio wave emissions. Radio waves can also give us information about an object's temperature, as well as the speed and direction of its movement. In addition, there are gamma ray telescopes, X-ray

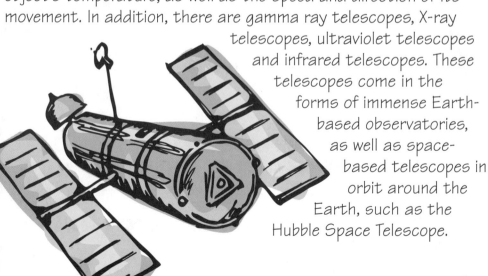

telescopes, ultraviolet telescopes and infrared telescopes. These telescopes come in the forms of immense Earth-based observatories, as well as space-based telescopes in orbit around the Earth, such as the Hubble Space Telescope.

All of
these tools,
in addition to our
endeavors in space
exploration, give us
volumes of information
about the
universe we live in.

SO WHAT EXACTLY HAVE WE LEARNED
ABOUT OUR UNIVERSE AFTER ALL THIS?

SO GLAD YOU ASKED! LET'S BEGIN.

CHAPTER 1:
THE BIG
BANG

And that which sings and contemplates in you is still dwelling within the bounds of that first moment which scattered the stars into space.

— **Kahlil Gibran**, *The Prophet*
[Alfred A Knopf, Random House, Inc. New York, 1951.]

In the beginning...
all things were one.

We refer to the original state of our universe as the **initial singularity**, wherein everything we know to exist - matter, energy, time, and space - existed as one single reality, unimaginably small and infinitely dense. This all-encompassing reality was smaller than an atom, and nothing else existed outside of it.

THAT SEEMS IMPOSSIBLE TO COMPREHEND!

If this notion seems impossible to comprehend - that's because it is! No one really understands and, in fact, when we trace the history of the universe back to this original state of affairs, the laws of science, as we know them,

essentially break down.

IT'S KIND OF CRAMPED IN HERE.

INITIAL SINGULARITY (SIZE GREATLY EXAGGERATED)

THE BIG

BANG!

THE UNIVERSE:

AH...MUC BETTE

But somehow - roughly 12 to 15 billion years ago - this initial singularity burst forth in what we call the **Big Bang**, giving birth to a creative, expanding, and dynamic universe!

Gravity is one of the fundamental forces by which matter is attracted to matter. On one hand, if the rate of expansion of the early universe had been much faster than it was, it would have been *too* fast for gravity to have been able to draw matter together. Neither galaxies, nor stars, nor planets, nor people could have ever existed, as all of the material unleashed by the Big Bang would have just scattered apart endlessly into infinity.

On the other hand, as **Stephen Hawking** states in *A Brief History of Time*...

"IF THE RATE OF EXPANSION ONE SECOND AFTER THE BIG BANG HAD BEEN SMALLER BY ONE PART IN A HUNDRED THOUSAND MILLION MILLION, THE UNIVERSE WOULD HAVE RE-COLLAPSED BEFORE IT EVER REACHED ITS PRESENT SIZE."

STEPHEN HAWKING

"I FEEL LIKE MY BRAIN IS SMALLER BY ONE PART IN A HUNDRED THOUSAND MILLION MILLION THAN STEPHEN HAWKING'S."

Fortunately for us, the rate of expansion after the Big Bang was apparently perfectly balanced with the amount of matter in the universe. If it hadn't been perfectly balanced, we wouldn't have the enormous and beautiful complexity that we see when we look up into the night sky - or, for that matter, when we look all around our world.

We still don't know for certain whether this is an . . .

open universe that will continue to expand forever into infinity, or if it is a . . .

closed universe that will one day re-collapse upon itself in a big crunch, to return to a state of singularity. But by most indications, we have several billion years before we even have to begin to worry about that.

17

Yes and no!

The Big Bang *is* a theory - but it's a theory supported by overwhelming evidence. The expansion of the universe, however, is not a theory, but something that can be observed. First and foremost, if the universe is expanding as we go forward in time then, if we trace the history of the universe backwards in time, we can visualize it all stemming from a common beginning.

Early in the 20th century, astronomers and physicists first exploring the idea of the Big Bang reasoned that, with so much matter and energy packed together, the early universe had to have been so hot, and so dense, that nuclear reactions would have occurred everywhere at once. Nuclear reactions release radiation in the form of radioactive particles so, if all of this is true, then we should still be able to detect the residual heat radiation of the Big Bang throughout the universe. By now, billions of years later, they estimated this radiation should have cooled immensely, to be just a few degrees above *absolute zero.*

ABSOLUTE ZERO IS THEORETICALLY THE LOWEST TEMPERATURE POSSIBLE.

EARLY IN THE 20TH CENTURY, ASTRONOMERS DID NOT YET HAVE THE TECHNOLOGY TO DETECT SUCH RADIATION. SO, FOR A TIME, THE THEORY OF THE BIG BANG SEEMED LIKE A GOOD ONE, BUT IT HAD YET TO BE CONFIRMED.

Decades later, in 1965, extremely sensitive radio telescopes were invented that were able to detect high frequency radio waves coming from beyond the solar system. These radio waves were isotropic - that is, they seemed to occur from every direction, in the same degree, all at once. Furthermore, the temperature of this radiation was about 2.7 degrees above absolute zero - incredibly close to the original theoretical estimates. This radiation is now known as the **cosmic microwave background radiation**, and is widely considered to be confirmation of the Big Bang event.

In the moments immediately after the Big Bang, the universe would have still been so tightly packed together that everything would have essentially existed as a single fireball of energy. But as the universe rapidly expanded, it would also cool, which would allow for different physical processes to emerge.

AS AN ANALOGY, THINK ABOUT HOW LIQUID WATER, WHEN COOLED AND FROZEN, TURNS INTO SOLID ICE.

The first forms to emerge within the first second after the Big Bang were **quarks** - the basic building blocks of elementary particles. Different combinations of these variously charged entities soon combined to form the subatomic particles of **protons** and **neutrons**. During the next three minutes, as the early universe continued to cool, these protons and neutrons then bonded together to form **atomic nuclei**.

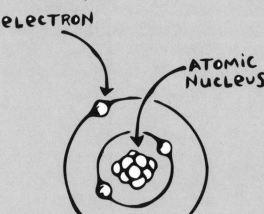

eLecTRON

ATOMIC NucLeuS

Although this occurred within just the first three minutes of time, it took hundreds of thousands of years before the universe had expanded and cooled enough for the subatomic particles of **electrons** to surround these atomic nuclei. And when the first electrons bound themselves into orbit around the first atomic nuclei, they were finally able to form the first **atoms** - the fundamental particles of matter.

IF THE CREATION OF MATTER OUT OF PURE ENERGY SEEMS DIFFICULT TO UNDERSTAND, LET ME TRY TO EXPLAIN.

IN MY **SPECIAL THEORY OF RELATIVITY**, I CAME UP WITH A LITTLE EQUATION THAT YOU MAY HAVE HEARD OF: $E = MC2$. TECHNICALLY, THIS MEANS THAT ENERGY (E) IS EQUAL TO MASS (M), TIMES THE SPEED OF LIGHT (C), MULTIPLIED BY ITSELF (SQUARED). BUT THIS IS REALLY JUST A FANCY WAY OF SAYING THAT MATTER AND ENERGY ARE TWO SIDES OF THE SAME COIN.

FOR EXAMPLE, IN THE NUCLEAR EXPLOSIONS THAT TAKE PLACE INSIDE STARS, TINY ATOMS ARE HEATED AT SUCH EXTREME TEM-PERATURES THAT THEY ARE FUSED TOGETHER TO FORM NEW ATOMS. BUT IN THE PROCESS, SMALL PARTS OF THE ORIGINAL ATOMS ARE BURNED OFF AND RELEASED AS TREMENDOUS AMOUNTS OF ENERGY, WHICH WE CAN SEE IN THE FORM OF LIGHT.

ON THE OTHER HAND, WHEN THE PURE ENERGY OF THE EARLY UNIVERSE EXPANDED AND COOLED, IT WAS TRANSFORMED INTO MATTER. SO, IN A NUTSHELL, WE CAN SAY THAT ENERGY IS EXTREMELY HOT MATTER, AND MATTER IS **FROZEN** ENERGY. WHICH IS PRETTY COOL!

21

Different combinations of protons, neutrons and electrons form different kinds of atoms, or elements. Roughly one billion years after the Big Bang, the simplest and most abundant elements -

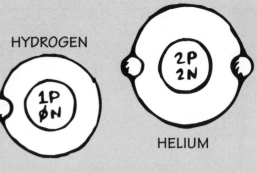

HYDROGEN

HELIUM

hydrogen and helium - began to condense, forming gaseous molecular clouds, or **nebulae** (Latin for 'clouds'). Eventually, through the force of gravity, the hydrogen and helium atoms within these molecular clouds collapsed upon themselves, forming the first stars.

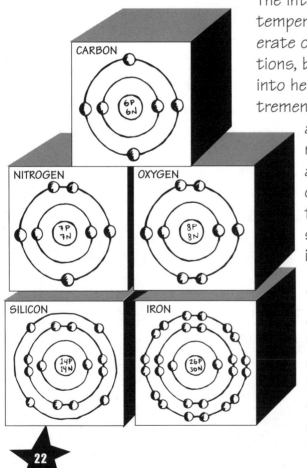

CARBON

NITROGEN

OXYGEN

SILICON

IRON

The intense pressures and temperatures within stars generate constant nuclear reactions, burning hydrogen atoms into helium, and processing tremendous amounts of energy and light. As stars reach old age, however - and what constitutes old age depends on the type of star - their systems become increasingly unstable.

This means that helium atoms become fused into heavier elements such as carbon, nitrogen, oxygen, silicon and iron. These elements (among many others) are the building blocks for planets and living things.

When a large, dying star finally explodes in what is called a 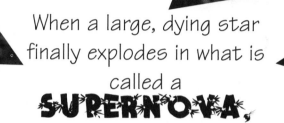 **SUPERNOVA,** these heavier elements are scattered into the universe - where they can then go on to build worlds like our own.

THE FANTASTIC FOUR

Physicists have been able to identify four fundamental forces in the universe that emerged from the Big Bang, and which act on various forms of matter.

The strongest of these forces is the **strong nuclear force**, which binds quarks together to create protons and neutrons and also binds protons and neutrons together to create atomic nuclei.

The next strongest force is the **electromagnetic force**, which is responsible for binding electrons around atomic nuclei to create atoms.

Less strong is the **weak interaction**, which is in charge of the natural disintegration of atomic nuclei, releasing radioactive energy in the form of subatomic particles.

Surprisingly, the weakest force in the universe is the force of **gravitation**, which attracts larger bodies of matter together, from collections of atoms to collections of galaxies.

While the other three forces are stronger than gravity, they can act only over very small distances. By contrast, gravity may be the weakest force, but its range is unlimited. For example, it may not be strong enough to assemble the constituent parts of a tiny atom, but it can help to keep the planets in orbit around the Sun, over the spaces of hundreds of millions of miles.

24

MORE ON THIS TO COME!

Still in the Dark

Of course, there is still much, much more about the nature of our universe that we still don't understand. For example, observations of many nearby galaxies, as well as our own, indicate that all of the visible matter known to exist may not be anywhere *near* the amount of matter *necessary* to explain how all of these vast systems are held together by gravity. This has led astronomers to believe that a mysterious, undetected source of matter - referred to as

Dark Matter

may account for as much as 90% of all of the matter in our universe.

More recently, observations of the expansion of our universe have also revealed another completely unexpected result: Whereas we would expect the gravitational effects of all of this matter in the universe to eventually *slow down* its rate of expansion, it instead seems as if the universe is currently expanding faster and faster! Once again, the precise nature of this *accelerating* force, now known as

Dark Energy,

remains a mystery. Nevertheless, it is currently believed to account for as much as two thirds of all of the energy in our universe!

CHAPTER 2:

THE SOLAR SYSTEM

Galaxies contain between one million and one trillion stars, and the universe is believed to contain as many as 100 billion galaxies. On an outer arm of a spiral galaxy we call the **Milky Way**, we find our solar system, which began its life in the form of a molecular cloud.

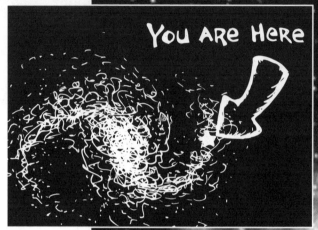

You Are Here

Around 5 billion years ago, the center of this *primordial* nebula began to collapse to form our star, the Sun. About five percent of the original nebulous material remained, however, continuing to revolve around this proto-star and eventually condensing into planets.

The French astronomer and mathematician **Pierre Simon, Marquis de Laplace** (1749 - 1827) and the German philosopher **Immanuel Kant** (1724 - 1804) both deserve credit for independently coming up with the theory of the origin and formation of the solar system, which, in its general terms, is most widely accepted today.

Immanuel Kant

The **Kant-Laplace nebular hypothesis** stated that - rather than being created as *is* - our solar system evolved from this earlier primordial state, a rotating nebula of gas and dust. They reasoned that the center of the solar nebula must have collapsed upon itself through the force of gravity to become our Sun, while the planets and their moons condensed from a surrounding disk, which originally looked something like the rings of Saturn.

Furthermore, it was because this primordial solar nebula originally rotated as one piece that the planets that emerged from it came to revolve around the Sun in the same direction in which the Sun rotates.

Pirre Simon, Marquis de Laplace

The Sun at the center of our solar system is an average star.

The temperatures at its center reach up to around 29,000,000 degrees Fahrenheit (16,000,000 degrees Celsius), with nuclear reactions constantly fusing its hydrogen atoms into helium, and processing energy at a rate equivalent to about 100 billion nuclear bombs per second!

It takes millions of years for this energy to escape the density and pressure of the Sun's core but, soon after it does, it fuels the entire solar system at the speed of light.

The Sun contains about 99% of all of the mass in our solar system, with a volume about 1,250,000 times the size of the Earth's. As the Earth and the other planets orbit the Sun, the Sun orbits the center of our galaxy at a speed of over 500,000 miles per hour (800,000 km/hr).

OH COME ON, DO I LOOK AVERAGE TO YOU?

*IT TAKES OUR PLANET **ONE YEAR** TO REVOLVE AROUND THE SUN, BUT IT TAKES THE SUN **OVER 250 MILLION YEARS** TO REVOLVE AROUND THE CENTER OF THE GALAXY.*

BEFORE YOU GO ANY FURTHER, ALLOW ME TO TELL YOU A LITTLE BIT MORE ABOUT HOW GRAVITY HOLDS THE SOLAR SYSTEM TOGETHER. YOU SEE, I HAD BEEN SITTING OUT IN MY APPLE ORCHARD ONE FINE ENGLISH AFTERNOON, WHEN I HAPPENED TO SPOT AN APPLE FALL FROM A TREE.

IT WAS THEN THAT I REALIZED THAT THE FORCE THAT PULLED THE APPLE DOWN TO THE GROUND WAS THE **SAME** FORCE THAT HOLDS THE PLANETS IN THEIR ORBITS AROUND THE SUN - WHICH IS WHY THEY DON'T GO FLYING OFF INTO SPACE!

HOWEVER, THE REASON WHY THE PLANETS DON'T GO CRASHING INTO THE SUN IS BECAUSE OF ANOTHER FACTOR INVOLVED, NAMELY, **MOMENTUM**, THAT IS, THE PRIOR STATE OF MOTION OF THE PLANETS, WHICH ACTS AGAINST THE FORCE OF THE SUN'S GRAVITY.

ALLOW ME TO ILLUSTRATE.

SAY WE WERE SOMEHOW TO STEP OUT INTO THE SOLAR SYSTEM AND GENTLY PLACE OUR APPLE WITHIN THE SUN'S GRAVITATIONAL PULL. WITHOUT ANY PRIOR SPEED ON THE APPLE'S PART, IT WOULD, IN TIME, BE PULLED STRAIGHT DOWN INTO THE SUN.

ON THE OTHER HAND, IF WE WERE TO **THROW** OUR APPLE OUT INTO THE SOLAR SYSTEM WITH TREMENDOUS SPEED, THEN THIS SPEED WOULD GIVE THE APPLE MOMENTUM AND DIRECTION, EVEN WHILE THE SUN'S GRAVITY WOULD KEEP IT IN ITS ORBIT - JUST AS IT DOES WITH ALL OF THE PLANETS IN OUR SOLAR SYSTEM.

MERCURY

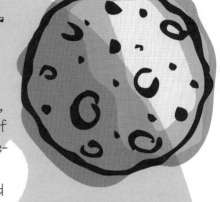

is the first planet from the Sun, whose rocky, moonlike surface reflects a bright white light on Earth, often making it brighter than most of the stars in our sky. Mercury's diameter is a little less than one-third the size of the Earth's, and is the second smallest planet in our solar system, after Pluto.

A year on Mercury goes by pretty fast. It only takes about 88 days for Mercury to circle the Sun, whereas it takes the Earth about 365 days. At the same time, a day on Mercury is pretty slow. Mercury rotates only once about every 58 days, whereas the Earth rotates once every 24 hours.

Because Mercury's rotation rate is so slow, and because it's so close to the Sun, at about 36 million miles (58 million km) away, it experiences huge changes in temperature. Mercury's daytime side, facing the Sun, reaches temperatures of up to 810 degrees F (430 C), while its nighttime side, facing away from the Sun, can dip down to -290 degrees F (-180 C).

"MERCURY IS NAMED FOR ME, THE ROMAN GOD OF TRAVEL, COMMERCE, THIEVERY AND CUNNING. AS THE MESSENGER OF THE GODS, I HAD WINGS ON MY SANDALS AND MY CAP, MAKING ME FAST AND SWIFT - JUST LIKE MY PLANET'S MOTION IN THE HEAVENS."

VENUS

is the second planet from the Sun, at about 67 million miles (108 million km) away. It takes about 225 days to orbit the Sun, and an even longer period of time - about 243 days - to rotate on its axis. Venus is the closest planet to the Earth, and is roughly the same size - just slightly smaller. Because its surface is almost completely covered by a layer of clouds, it reflects about 80% of the light from the Sun, making it the brightest of all of the planets in our sky. Depending on its position, it can appear as much as twelve times brighter than the brightest star, Sirius.

Beneath its layer of clouds - made out of concentrated sulfuric acid - Venus is a hot, volcanically active world. Around 500 million years ago, Venus is believed to have gone through an extraordinarily dramatic volcanic period, which covered its surface with floods of lava, beneath an atmosphere made up almost entirely of carbon dioxide (CO_2). Carbon dioxide, known as a **greenhouse gas**, has the effect of trapping the Sun's heat in the atmosphere, creating an average, steady surface temperature on Venus of about 858 degrees F (459 degrees C).

Depending on its position in relation to the Sun, the planet Venus is sometimes seen in the morning and sometimes in the evening. The ancients, however, were unaware that these two exceptionally bright objects were actually one and the same! The morning star they referred to as Lucifer, the light-bearer, who was thought to bring in the day, while the evening star was Venus herself, the Roman goddess of love and beauty.

Our

EARTH

is the third planet from the
Sun, at about 93 million
miles (150 million km) away.
The average distance from
the Earth to the Sun is also
known as the **astronomical
unit (AU)**, and is used as a measure of other astronomical distances.

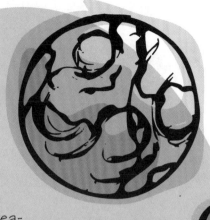

MOON

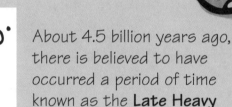

About 4.5 billion years ago,
there is believed to have
occurred a period of time
known as the **Late Heavy
Bombardment**, in which frag-
ments of the early solar system systematically impacted the
planet. One of the fragments to collide with the young Earth was
very likely a rogue planet, or **planetesimal**, at least the size of
Mars. At that geological time, the Earth would have still been a
largely molten planet beneath its thin outer crust. As a result, a
collision of this scale would have been so intense that significant
parts of the Earth's crust and mantle would have been ejected
into space upon impact. The debris of molten rock would have
then coalesced and solidified
to form
our **Moon**,
developing
an orbit
very close
to the
Earth.

SINCE THEN, THE MOON HAS SLOWLY DRIFTED AWAY FROM THE EARTH, AND IT CONTINUES TO DO SO TODAY AT THE RATE OF ABOUT ONE AND A HALF INCHES PER YEAR.

The chaotic period of the Late Heavy Bombardment - in which the solar system may have resembled a giant game of billiards - would also neatly explain a number of other oddities.

FOR EXAMPLE: because the solar system originally emerged from a single rotating nebula, we would naturally expect all of the planets in the solar system to revolve around the Sun in the same direction in which the Sun rotates - and this is indeed the case.

We would also generally expect all of the planets to rotate on their axes in the same direction, which most, but not all, of them do.

FOR EXAMPLE: Venus rotates on its axis in a direction *opposite* to that of the other planets, while Uranus rotates at a 90-degree tilt to the plane of its orbit. These planets are therefore exceptions to the general rule.

MOON ☽

BUT if they were once knocked off of their regular rotations by rogue planetesimals, then their quirky behavior is much easier to understand.

Just past the Earth's orbit, we find 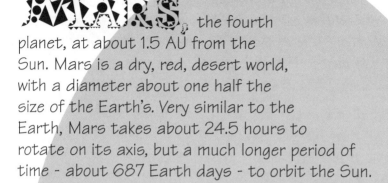 the fourth planet, at about 1.5 AU from the Sun. Mars is a dry, red, desert world, with a diameter about one half the size of the Earth's. Very similar to the Earth, Mars takes about 24.5 hours to rotate on its axis, but a much longer period of time - about 687 Earth days - to orbit the Sun.

PHOBOS

DEIMOS

> REMEMBER, AU STANDS FOR ASTRONOMICAL UNIT, WHICH IS THE AVERAGE DISTANCE BETWEEN THE EARTH AND THE SUN.

Although it now appears to be a dead, rocky planet, it was once volcanically active and boasts the largest (but extinct) volcano in the known solar system: **Olympus Mons**, 370 miles (595 km) across and 15 miles (24 km) high - about three times as high as Mount Everest! Mars also has two polar ice caps, and small amounts of water vapor and oxygen in its atmosphere, which is otherwise composed mainly of carbon dioxide. The two small, oddly-shaped moons of Mars, **Phobos**, at about 13 miles (20 km) in diameter, and **Deimos**, about 7.5 miles (12 km) in diameter, may have originally been asteroids that were long ago trapped in its orbit.

> THE RED PLANET WAS NAMED FOR ME, THE ROMAN GOD OF WAR, WHILE ITS TWO MOONS WERE NAMED AFTER MY TWO ATTENDANT SONS, PHOBOS, OR FEAR (FROM WHOSE NAME WE GET THE WORD 'PHOBIA'), AND DEIMOS, OR PANIC.

ASTEROIDS,

also called **minor planets**, are smaller, rocky objects, probably left over from the tumultuous beginning of the early solar system. Most of the asteroids in our solar system vary from less than a mile to roughly 600 miles (1000 km) in diameter, and are found in regular orbits between Mars and Jupiter - at about 2.3 AU from the Sun.

Although most asteroids are well behaved, staying right where they belong, some asteroids have highly elliptical orbits that will occasionally bring them near the Earth. From time to time, the Earth has been hit by asteroids that intersect our orbit and enter our atmosphere. Most of these have been small and relatively harmless. However, it is estimated that as many as 2,000 asteroids more than a half a mile in diameter could possibly collide with the Earth at times far into the future.

One asteroid, about 6 miles (10 kilometers) in diameter that collided with the Earth about 65 million years ago, is now widely believed responsible for the extinction of the dinosaurs. The explosion from such a gigantic impact could have easily blanketed the Earth's atmosphere with trillions of tons of debris - enough to block out the

THERE GOES THE NEIGHBORHOOD

light and the heat from the Sun for many months on end. That would then result in a planet-wide ice age. As evidence, the geological record reflects exactly such a level of debris and, in 1997, an impact crater was discovered in Mexico's Yucatan Peninsula. This almost certainly confirms this dinosaur extinction theory.

Fortunately for us, catastrophic collisions like this one probably only occur about once every 100 million years or so!

Just past the asteroid belt is

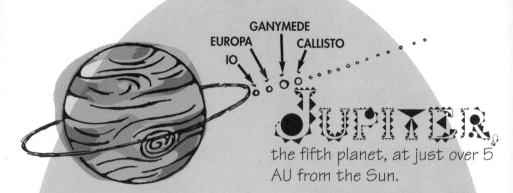

the fifth planet, at just over 5 AU from the Sun.

Jupiter is the largest planet in the solar system, with a volume about 1,400 times the size of the Earth's.

Unlike the smaller, rocky, inner planets (Mercury, Venus, Earth and Mars), the larger, outer planets are mostly made of gas. Astronomers believe that Jupiter has a small core made out of solid rock and iron. This core is surrounded by an ocean of liquid metallic hydrogen, which in turn is surrounded by an immense gaseous atmosphere, primarily composed of hydrogen and helium - the same elements that compose stars. In fact, Jupiter's composition is probably directly left over from the early primordial nebula that formed the Sun. Actually, if Jupiter had been about 80 times larger, it would have been large enough to generate nuclear reactions, thus becoming the second star in our solar system!

Jupiter takes almost 12 years to orbit the Sun, but only about 10 hours to rotate—an. exceptionally fast rotational period. That's why the planet bulges in the middle, which can be seen clearly when viewed through a good telescope.

YOU'VE GOT TO STOP SPINNING SO FAST OR YOU'RE NEVER GOING TO GET RID OF THAT GUT!!

Jupiter's light and dark bands are atmospheric currents, constantly swirling clouds of frozen ammonia and methane. The outer, gaseous surface of Jupiter is turbulent - lots of high winds, strong lightning, and wild cyclone-like storms.

Jupiter's distinctive Great Red Spot is a cyclone twice the size of Earth! It's the largest storm in the solar system, with winds reaching up to 270 miles (435 km) per hour. It has raged for at least three centuries!

Jupiter is surrounded by a small ring of solid, rocky particles that probably came from the nearby asteroid belt.

Jupiter has at least 16 moons, four of which are about the size of the planet Mercury! - They can be seen easily with a standard pair of binoculars. These four moons are called the *Galilean satellites*, after **Galileo Galilei** (1564 - 1642), the Italian astronomer who first discovered them in the year 1610.

Io is a rocky world, alive with fierce volcanic activity.

Europa has a thin atmosphere of oxygen and an icy exterior, beneath which may lay an immense global ocean - and a distant possibility for life!

Ganymede and **Callisto**, the two largest moons, are the outermost of the four Galilean satellites, and are both icy worlds covered with craters.

JUPITER, THE KING OF THE PLANETS, WAS NAMED AFTER ME, THE ROMAN KING OF THE GODS, WHILE MOST OF THE MOONS THAT ORBIT JUPITER ARE NAMED AFTER THE WOMEN IN MYTHOLOGY WHOM I PURSUED.

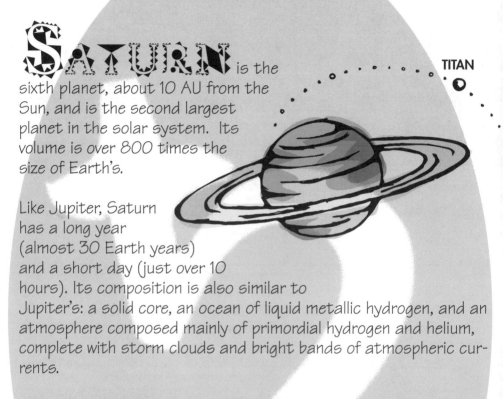

SATURN is the
sixth planet, about 10 AU from the
Sun, and is the second largest
planet in the solar system. Its
volume is over 800 times the
size of Earth's.

TITAN

Like Jupiter, Saturn
has a long year
(almost 30 Earth years)
and a short day (just over 10
hours). Its composition is also similar to
Jupiter's: a solid core, an ocean of liquid metallic hydrogen, and an
atmosphere composed mainly of primordial hydrogen and helium,
complete with storm clouds and bright bands of atmospheric currents.

Saturn's distinctive bright rings, made up of frozen gas, rock, and
ice, were also discovered by Galileo in 1610, and can also be seen
clearly with a good telescope. Saturn is furthermore surrounded
by at least 20 moons. The most interesting of these is the
largest moon, **Titan**, a world larger than Mercury and almost as
large as Mars. Titan has a thick atmosphere, which like our own is
composed mostly of nitrogen. Beneath its cloudy exterior, its
interior is believed to be made of rock and ice and, like Europa,
there is also a possibility that Titan might be capable of harbor-
ing some form of life.

"THE SECOND LARGEST PLANET WAS NAMED AFTER ME, A TITAN NAMED SATURN. I WAS THE FATHER OF THE GODS AND THE KING OF THE UNIVERSE - UNTIL I WAS DETHRONED AND SURPASSED IN GREATNESS BY MY SON, JUPITER. MANY OF SATURN'S MOONS, SUCH AS ATLAS, RHEA, AND HYPERION, ARE NAMED AFTER OTHER TITANS, WHILE MOST OF THE REST REPRESENT A VARIOUS AND SUNDRY COLLECTION OF MYTHOLOGICAL CHARACTERS."

The seventh planet,

URANUS,

is about 19 AU from the Sun. Like the other gas giants, Uranus is largely composed of hydrogen and helium, but it gets its bluish-green color from significant amounts of methane in its atmosphere as well. Uranus is the third largest planet in the solar system, with a volume about 64 times the size of Earth's. Uranus is also surrounded by nine faint rings of dust, rock, and ice, as well as 15 small moons. It has a 17-hour day and takes 84 Earth years to orbit the Sun.

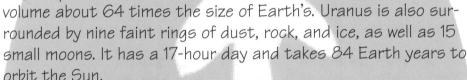

Before the Titans gave birth to the gods, the Earth and the Sky gave birth to the Titans. The Greeks called Mother Earth, Gaea, and Father Sky, Uranus - after whom the seventh planet is named. Unlike the rest of the planets, the moons of Uranus are not named after figures from mythology, but rather after characters from Shakespeare's plays.

Uranus was first discovered in 1781 by the British astronomer, **Sir William Herschel** (1738 - 1822), and for a time it was actually called the planet Herschel (which is probably a nicer name!).

"I'LL NEVER UNDERSTAND WHY THEY CHANGED THE NAME!"

Sir William Herschel

The eighth planet,

NEPTUNE,

is about 30 AU from the Sun, and is very similar in size, color, and composition to Uranus. The fourth largest planet, its volume is about 58 times the size of the Earth's. Its atmosphere is also composed largely of hydrogen and helium, and it gets its bright blue color from methane as well. Storms are also raging on Neptune's surface, with winds reaching up to 1,500 miles (2400 km) per hour - the strongest winds known in the solar system! Neptune has a 16-hour day and takes 165 Earth years to orbit the Sun. It has five small, dusty rings, and eight moons, the largest of which, **Triton**, is another intriguing icy world, with a nitrogen atmosphere, active geysers, and perhaps a possibility for life.

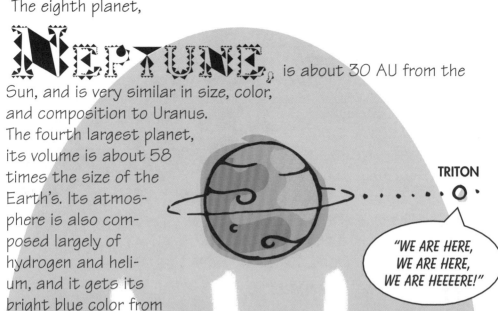

TRITON

"WE ARE HERE, WE ARE HERE, WE ARE HEEEERE!"

The bright blue planet of Neptune is named after the Roman god of the sea. Many of Neptune's moons, such as **Triton** and **Proteus**, are named after other sea divinities, while the moons of **Naiad** and **Nereid** are named after different types of water dwelling nymphs.

The ninth planet,

is about 39 AU from the
Sun. Pluto is a yellowish,
rocky world with a light
methane atmosphere. It
is also the smallest planet, with a diameter about
one fifth the size of the Earth's -

CHARON

or about two-thirds the size of our Moon. It takes over six days
to rotate, and almost 250 Earth years to orbit the Sun. Pluto
also has one moon, **Charon**, which is very close in size to Pluto
itself.

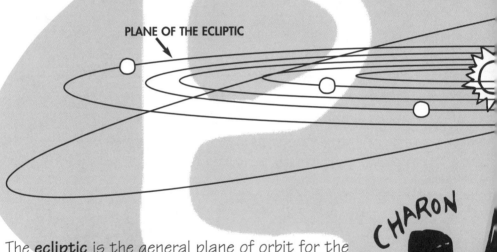

PLANE OF THE ECLIPTIC

The **ecliptic** is the general plane of orbit for the
Earth and all of the other planets in our solar system - except for Pluto, whose orbit tilts away from
the plane of the ecliptic at an angle of about 17
degrees. Furthermore, Pluto's highly elliptical orbit
sometimes actually brings it closer to the Sun
than Neptune.

CHARON

Because Pluto is so small, and because its orbit is so eccentric, some astronomers question whether it really deserves to be considered a planet at all! Between 30 and 100 AU from the Sun, on the outer edge of the known solar system, lies the **Kuiper belt** (rhymes with *piper*), a ring of rocky, icy bodies, much smaller than most planets and much larger than most asteroids. If Pluto isn't considered a planet, it could someday be turned into a Kuiper object (and its little moon too!). But until that officially happens, Pluto is still the ninth planet - just a highly unusual one!

PLUTO'S ORBIT

Pluto, the farthest planet from the Sun, is named after the Roman god of the dead, while its companion, Charon, is named after the ferryman who took the souls of the dead across the river of lamentation, and into the underworld.

VIEWING THE PLANETS

The planets can be distinguished by their steady light (whereas stars twinkle), their locations along the path of the ecliptic (the same path that the Sun takes across the sky), and by their wandering nature in relation to the stars. In fact, the word 'planet' comes from the Greek word *planetos*, meaning 'wanderer'.

AND MANY LOCAL NEWSPAPERS WILL OFTEN TELL YOU EXACTLY WHEN THE MAJOR PLANETS RISE AND SET ON ANY GIVEN NIGHT.

The Evening Herald

PLANETS RISE TO THE OCCASION!

GREATEST EASTERN ELONGATION

During the course of a single night, as the Earth turns, the planets appear to move from east to west, just like the Sun, the Moon, and the stars.

But a closer look reveals that from one night to the next, against the background of the stars, the planets generally tend to move from

west to east.

The major planets - Mercury, Venus, Mars, Jupiter, and Saturn - can all be seen easily with the naked eye. Uranus, however, is quite difficult to see with the naked eye, and Neptune and Pluto can only be seen with a good telescope.

While all of the planets wander the skies at various times throughout the year, the best times to view the inner planets, Mercury and Venus, are when they are at or near their greatest **elongations**. At such times, from our perspective, they are at their farthest points away from the Sun and its glare. At their greatest *western* elongations, they can be seen just after sunset, and at their greatest *eastern* elongations, they can be seen just before sunrise.

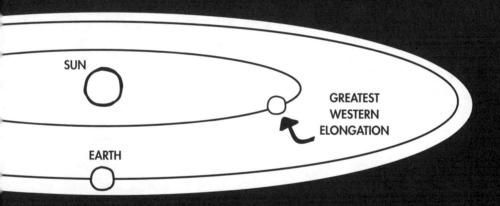

Mercury reaches its greatest elongations between three to five times a year. When it does, it appears just above the horizon at a little under two hours before sunrise or after sunset.

With its longer orbit, Venus can reach elongation, at most, twice a year, and some years not at all. When Venus reaches greatest elongation, it appears well above the horizon about three hours before sunrise or after sunset.

The best times to view the outer planets are when they are at **opposition**.

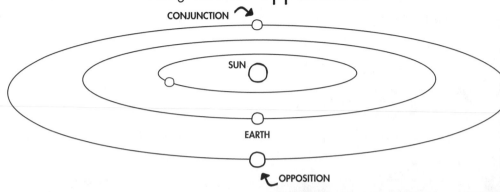

CONJUNCTION

SUN

EARTH

OPPOSITION

At such times, they are 180 degrees *opposite* from the Sun in our sky. Thus, when the Sun sets, the planet in opposition will rise in the east and will remain in the sky throughout the night, at its closest and brightest point to the Earth.

MARS IS IN OPPOSITION ABOUT ONCE EVERY TWO YEARS; JUPITER, ABOUT ONCE EVERY THIRTEEN MONTHS; SATURN, URANUS, NEPTUNE, AND PLUTO, ROUGHLY ONCE A YEAR.

THE SEARCH FOR LIFE ON OTHER WORLDS

So far, the Earth is the only planet in the solar system and, indeed, the only planet in the universe in which life is known to exist. Biologists believe that life originated on Earth, over 3.5 billion years ago, in the form of single-celled microorganisms, which somehow acquired the abilities to process energy, reproduce, and later to evolve. The ingredients for life - which don't appear to be too uncommon elsewhere in the universe - seem to include organic compounds (which are molecules based on carbon atoms), water, and some form of energy, such as light or heat. But although on one level it's easy enough to say *what* life is (We know it when we see it!), on another level it is still not understood *how* or *why* life originated in the first place. When millions of non-living organic compounds somehow organize themselves and spring to life, this is no small source of wonder!

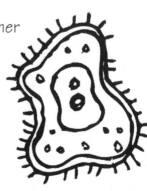

That said, the argument in favor of life beyond Earth goes something like this: The universe in which we live contains a countless number of stars, and many distant solar systems, not too unlike our own, have already been observed elsewhere in the galaxy. In such a vast and complex place, with so many possibilities, it shouldn't be too surprising if life on other worlds turned out to be somewhat common.

The more skeptical argument, however, asserts that life itself is such a singular mystery that, for all we know, it may have only originated at one time, and on one planet - which just so happens to be our own.

As yet, we have no conclusive evidence that life exists anyplace else but home.

A famous Martian meteorite, which NASA went public with in 1996, showed signs that some interpreted as evidence that microorganisms may have once existed on Mars, but in the final analysis it was a long way off from proof.

Nevertheless, absence of evidence isn't evidence of absence, as the saying goes, and hasn't stopped scientists from
SEARCHING.

Some scientists reason that if life *were* able to arise on other planets, and evolve into creatures with advanced forms of intelligence, then perhaps these extraterrestrial beings would be able to transmit communication through space via radio waves, just as we do with radio and television.

Radio waves may not be visible to the human eye, but just like waves of visible light, they're strong enough to travel vast distances at top speed.

While a standard optical telescope pointed at a star is designed to receive that star's visible light, a *radio telescope* pointed at a star is designed to receive that star's radio wave emissions. And while stars and other objects in the universe do emit radio waves naturally, an artificial message sent via radio waves should appear to be quite distinct.

THE ARECIBO RADIO TELESCOPE
IS THE WORLD'S LARGEST RADIO TELESCOPE.

IT'S IN PUERTO RICO. IT'S 305 METRES (1000 FEET).
SETI GETS TO USE IT FOR ABOUT 3 WEEKS EVERY
SPRING AND ANOTHER 3 WEEKS EVERY FALL.

Since the 1960's, there have been repeated attempts to search the heavens for radio wave messages from other stars. The most famous of these has been the Search for Extra Terrestrial Intelligence (SETI). This program was officially begun by NASA in 1992, only to have its budget completely cut by the US Congress in 1993. Nevertheless, it continues today through the support of private funding.

So far, no messages have been found, but then again this isn't simple work. In order to receive a possible signal, a radio telescope has to be pointed at a specific star - and there happen to be quite a lot to choose from! SETI's goal is to receive and analyze data from 1,000 stars in the galaxy. Furthermore, to assist in this endeavor, the University of California at Berkeley has a program called SETI@Home, which allows individuals to participate in this project over the Internet. After downloading free software, a personal computer can be used to help analyze radio wave data - which could just end up being the first signals from intelligent life beyond our planet! One way or another, the search will go on...

CHAPTER 3:
EARTH, SUN, MOON

As the Sun revolves around the Milky Way Galaxy, the Earth revolves around the Sun, and the Moon revolves around the Earth.

The relationship between the Earth and the Sun defines the Earth's seasons and the length of the year,

and the relationship between the Earth, the Sun, and the Moon creates some rather remarkable celestial events.

LET'S TAKE A CLOSER LOOK...

THE EARTH'S ORBIT

Cruising at about 67,000 miles per hour (108,000 km/hr), it takes our planet one year to complete its roughly circular, counterclockwise orbit around the Sun.

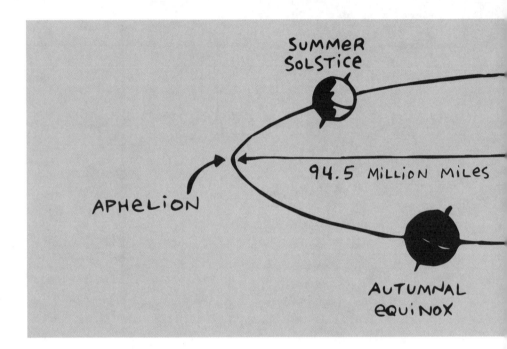

To be more precise,

it takes 365.2422 days, which is to say, 365 days, 5 hours, 49 minutes, and 12 seconds! We try to ignore this extra 24% of a day in our calendars until about once every four years, when we just can't put it off any longer. By that time, an entire extra day has added up, so we add February 29th into the leap year, to try to keep our calendars roughly up to speed with the solar year.

To be even more precise,

our orbit around the Sun isn't exactly circular. Technically, it's an ellipse. The Earth's average distance from the Sun is about 93 million miles. However, each year, around January 4th, we reach our **perihelion**, our closest point to the Sun, when we're about 91.5 million miles away, and six months later, around July 4th, we reach our **aphelion**, our farthest point from the Sun, when we're about 94.5 million miles away.

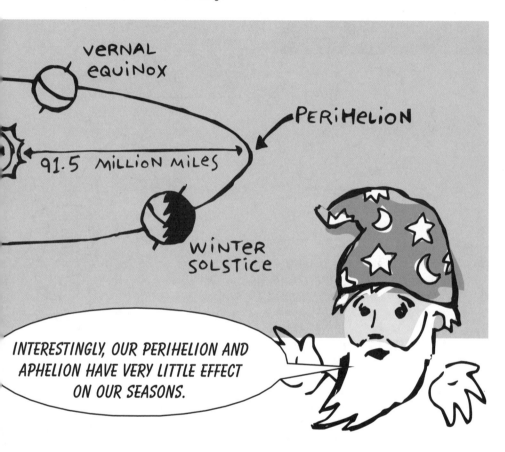

When the Southern Hemisphere experiences summer at perihelion, it receives just 6% more solar energy than when the Northern Hemisphere experiences summer at aphelion. Our seasons are determined instead by the relationship between the tilt of the Earth's axis of rotation and the degree of the Sun's most direct heat, that is, the plane of the ecliptic.

KEPLER'S LAWS

Our distance from the Sun doesn't determine our seasons, but it does determine our speed. This was first discovered by the German astronomer, **Johannes Kepler** (1571 - 1630), in the early 17th century...

MY *FIRST LAW* OF PLANETARY MOTION STATED THAT ALL PLANETS WILL ORBIT THEIR SUN IN ELLIPTICAL PATHS, WITH THEIR SUN AS ONE FOCUS OF THEIR ORBIT - THAT IS, CLOSER TO ONE END OF THEIR ORBIT THAN THE OTHER.

ELLIPTICAL ORBIT

PLANET

FOCUS 1   FOCUS 2

MY *SECOND LAW* OF PLANETARY MOTION WAS THAT AN ELLIPTICAL PATH WILL CAUSE VARIATIONS IN SPEED. THUS, WHEN PERIHELION BRINGS US CLOSER TO THE SUN, THE EARTH WILL MOVE A LITTLE FASTER ALONG IN ITS ORBIT, AND WHEN APHELION TAKES US FARTHER FROM THE SUN, THE EARTH WILL MOVE A LITTLE MORE SLOWLY ALONG IN ITS ORBIT.

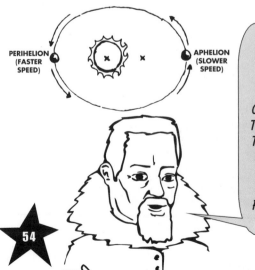

PERIHELION (FASTER SPEED)

APHELION (SLOWER SPEED)

SIMILARLY, MY *THIRD LAW* OF PLANETARY MOTION WAS THAT A PLANET'S DISTANCE FROM ITS SUN WILL ALSO DETERMINE HOW FAST IT GENERALLY MOVES IN ITS ORBIT. IN OTHER WORDS, THE CLOSER THE PLANET IS TO THE SUN, THE FASTER IT WILL MOVE; THE FARTHER A PLANET IS FROM THE SUN, THE SLOWER IT WILL MOVE. THIS IS ONE REASON WHY MERCURY - SO CLOSE TO THE SUN - HAS AN ORBIT OF JUST 88 DAYS, WHILE PLUTO - SO FAR AWAY FROM THE SUN - HAS AN ORBIT OF NEARLY 250 EARTH YEARS.

The **ecliptic** is the plane of our orbit around the Sun and, as we have already mentioned, it is also the general plane of orbit for all of the other planets in our solar system (except for that wacky, lovable Pluto!). One way to conceptualize the plane of the ecliptic is to imagine a line connecting the center of the Sun to the center of the Earth. As the Earth revolves around the Sun, our imaginary line draws a plane in its path. Extend this plane out into space, beyond the Earth, and you have the plane of the ecliptic.

THIS IS HOW WE CAN SEE THE PATH OF THE ECLIPTIC FROM EARTH.

EAST

SOUTH

WEST

NORTH

From our perspective on Earth, we can also conceptualize the *path* of the ecliptic as the path of the Sun against the back-ground of the sky. But, of course, the Sun doesn't actually move across the sky. It just appears that way because the Earth is spinning on its axis.

The Earth's **axis** can be thought of as a straight line going directly through the cen-ter of the Earth, with one end at the North Pole, and the other end at the South Pole. However, in relation to the plane of the ecliptic, the Earth's axis is tilted at about 23.5 degrees.

N

S

VERNAL EQUINOX

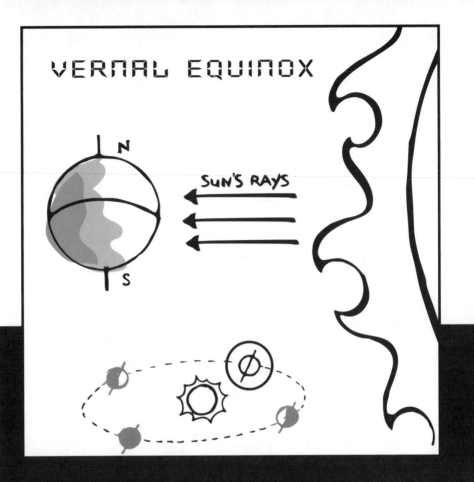

The tilt of the Earth's axis of rotation, and its changing relationship to the plane of our orbit, naturally divides the year into four seasons. On the **vernal equinox**, which falls around March 21st, the Earth's axis of rotation is at a 23.5 degree sideways tilt in relation to the plane of the ecliptic (rather than being tilted northward or southward), so that the Sun's light covers both the Northern Hemisphere and Southern Hemisphere equally. At noon, the Sun is directly above the **equator**, at zero degrees latitude, and all over the world the day and the night are equally about 12 hours long.

As the Earth spins on its axis, the most direct sunlight (also called *solar noon*) remains directly above the equator. Because the Sun remains above the equator, from our perspective during the day, it will appear to rise in the east, and set in the west.

In the three months following the vernal equinox, the Earth will complete the first quarter of its revolution around the Sun, moving 90 degrees, counterclockwise, in its 360-degree orbit. During this time, the path of the Sun will slowly move north, from being directly above the equator at noon, to being directly above the tropic of Cancer, at 23.5 degrees north latitude.

In the Northern Hemisphere, the days will become **longer** and the nights will become shorter.
In the Southern Hemisphere, the days will become shorter and the nights will become **longer**.

SUMMER SOLSTICE

SUN'S RAYS

N

S

On the **summer solstice**, which falls around June 21st, the North Pole is tilted sunward, 23.5 degrees, and the Sun's rays beat down most directly upon the Northern Hemisphere. At noon, the Sun is directly above its north-ernmost point, the **tropic of Cancer**, at 23.5 degrees north latitude. In the Northern Hemisphere, this is the longest day and the shortest night of the year. In the Arctic Circle, above 66.5 degrees north latitude, there are 24 hours of daylight!

At the same time, the Southern Hemisphere is experiencing winter, because the South Pole is tilted away from the Sun. For the Southern Hemisphere, this is the winter solstice - the shortest day and the longest night of the year. In the Antarctic Circle, below 66.5 degrees south latitude, there are 24 hours of darkness.

In the three months following the summer solstice, the Earth will complete the second quarter of its revolution around the Sun, moving another 90 degrees, counterclockwise, in its 360-degree orbit. During this time, the path of the Sun will slowly move south, from being directly above the tropic of Cancer at noon, to being directly above the equator.

At this time, as the Earth spins on its axis, the most direct sunlight remains directly above the tropic of Cancer. Because the Sun remains at its northernmost point, from our perspective during the day, it will appear to rise in the northeast, and set in the northwest.

In the Northern Hemisphere, the days will become shorter and the nights will become longer.
In the Southern Hemisphere, the days will become longer and the nights will become shorter.

AUTUMNAL EQUINOX

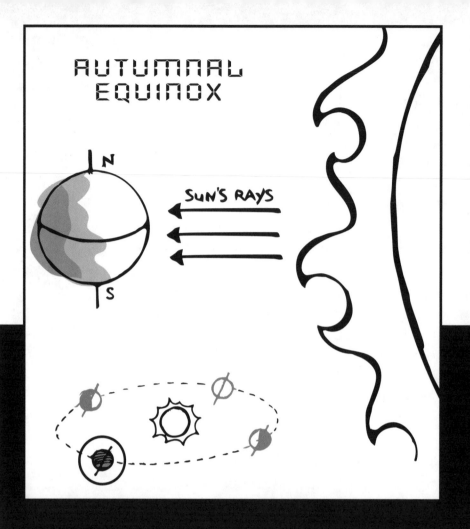

SUN'S RAYS

By the **autumnal equinox**, which falls around September 22nd, the Earth's axis of rotation has again reached a 23.5 degree sideways tilt in relation to the ecliptic, so that the Sun's light covers both the Northern Hemisphere and the Southern Hemisphere equally. At noon, the Sun is once again directly above the equator, and all over the world the day and the night are equally about 12 hours long.

As the Earth spins on its axis, the most direct sunlight from the ecliptic again remains directly above the equator. From our perspective during the day, the Sun will again appear to rise in the east, and set in the west.

In the three months following the autumnal equinox, the Earth will complete the third quarter of its revolution, moving another 90 degrees, counterclockwise, along its 360-degree orbit. During this time, the path of the Sun will continue to move south, from being directly above the equator at noon, to being directly above the tropic of Capricorn, at 23.5 degrees south latitude.

In the Northern Hemisphere, the days will continue to become shorter and the nights will continue to become longer. In the Southern Hemisphere, the days will continue to become longer and the nights will continue to become shorter.

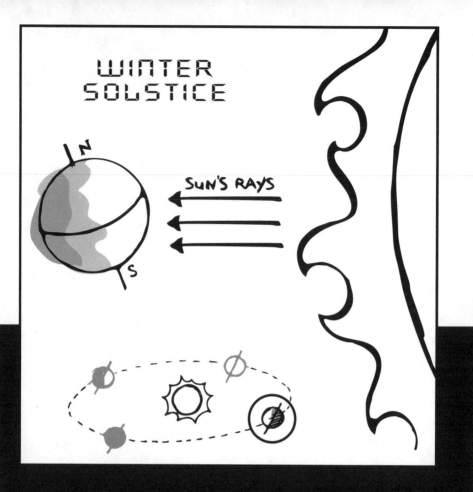

WINTER SOLSTICE

SUN'S RAYS

On the **winter solstice**, which falls around December 22nd, the Earth's South Pole is tilted sunward, 23.5 degrees, and the Sun's rays beat down most directly upon the Southern Hemisphere. At noon, the Sun is directly above its southernmost point, the **tropic of Capricorn**, at 23.5 degrees south latitude. In the Southern Hemisphere, this is the summer solstice, the longest day and the shortest night of the year. In the Antarctic Circle, below 66.5 degrees south latitude, there are 24 hours of daylight.

At this time, as the Earth spins on its axis, the most direct sunlight from the ecliptic remains directly above the tropic of Capricorn. From our perspective during the day, the Sun will appear to rise in the southeast, and set in the southwest.

At the same time as the Southern Hemisphere is experiencing summer, the Northern Hemisphere is experiencing winter. In the Northern Hemisphere, the winter solstice is the shortest day and the longest night of the year. In the Arctic Circle, above 66.5 degrees north latitude, there are 24 hours of darkness.

In the three months following the winter solstice, the Earth will complete the final quarter of its revolution around the Sun, moving another 90 degrees, counterclockwise, to complete the 360-degree orbit that it began at the last vernal equinox. During this time, the path of the Sun will slowly move north again, from being directly above the tropic of Capricorn, to being back above the equator.

In the Northern Hemisphere, the days will become **longer** and the nights will become shorter.

In the Southern Hemisphere, the days will become shorter and the nights will become **longer.**

63

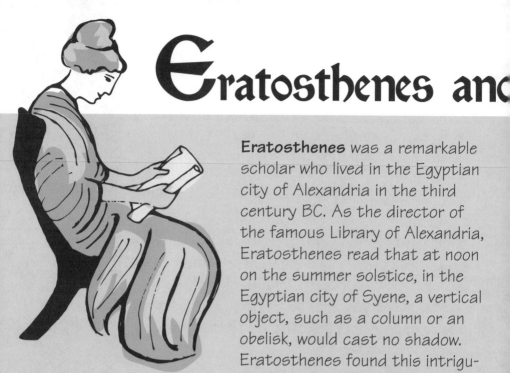

Eratosthenes and

Eratosthenes was a remarkable scholar who lived in the Egyptian city of Alexandria in the third century BC. As the director of the famous Library of Alexandria, Eratosthenes read that at noon on the summer solstice, in the Egyptian city of Syene, a vertical object, such as a column or an obelisk, would cast no shadow. Eratosthenes found this intriguing because in Alexandria, which was about 500 miles (or 800 kilometers) to the north of Syene, vertical objects did in fact cast shadows at this time.

As it happened, Syene lay essentially right on top of the Tropic of Cancer, at about 23.5 degrees north latitude. Thus, at noon on the summer solstice, the Sun was pretty much directly overhead in Syene. But because Alexandria was further to the north - and because the Earth is curved - this would not have been the case in Alexandria.

he Summer Solstice

Incredibly, Eratosthenes was able to figure out that this was indeed the explanation - which meant that the Earth wasn't flat, as many people at the time believed.

Eratosthenes went on to compare the difference between the directions of the shadows in Alexandria and Syene, and found that they differed by about 7 degrees. From this, he was able to deduce that between the two cities the Earth must have had a curvature of about 7 degrees. If there were 360 degrees of curvature around the edge of a sphere, and if 360 divided by 7 yielded 51, then, he reasoned, the distance between Alexandria and Syene must have been very nearly one-fiftieth of the distance all the way around the Earth's surface.

As a result, not only did Eratosthenes figure out that the world was round, but by multiplying the number of miles between Alexandria and Syene by 51, he was also able to figure out, very closely, just how big the planet actually was.

THE MOON

Just as the Earth has a counterclockwise, slightly elliptical orbit around the Sun, so does the Moon have a counterclockwise, slightly elliptical orbit around the Earth. When the Moon is at its **perigee**, its closest point to the Earth, it's about 226,000 miles away, about 28.5 times the length of the Earth. When the Moon is at its **apogee**, its farthest point from the Earth, it's about 252,000 miles away, about 32 times the length of the Earth. At perigee, the Moon appears about 12% larger than at apogee.

EARTH N 252,0

225,000 MILES

MOON

PERIGEE

Compared to the size of the Earth, the Moon is relatively small, with a radius just about one-fourth the size of the Earth's. If you were to get in your moon-buggy and drive from one side of the Moon to the other, it would be about the same distance as driving across the United States, say from Boston to San Francisco. Nevertheless, it is still large enough to have a considerable effect on the ocean's tides, which regularly rise and fall in accord with the Moon's gravitational pull. High tide will always occur shortly after the Moon reaches its **zenith**, its highest position overhead.

MOON

EARTH

Just as one half of the Earth is always lit by the Sun, so one half of the Moon is also always lit by the Sun. However, from our perspective, as the Moon revolves around the Earth, it appears to pass through several different phases.

At the **new moon** phase, the side facing away from the Earth is lit, and from our perspective, the Moon is shrouded in darkness. As the Moon revolves, it enters the **waxing crescent** phase, and each night a larger crescent becomes visible. By the time it enters the **first quarter** phase, a half moon can be seen. As it continues to revolve, it enters the **waxing gibbous** phase, still appearing to become larger and larger until it becomes a **full moon**, when the entire lit side can be seen from Earth. Continuing, it enters the **waning gibbous** phase, and less and less of it is seen each night. At the **last quarter** phase, once again, a half moon is seen. Finally, during the **waning crescent** phase, each night a smaller crescent is visible, until the Moon once again enters the new moon phase.

APOGEE
MOON
es

The **lunar month** is the period of time it takes for the Moon to complete this cycle. Traditionally, the lunar month is reckoned from new moon to new moon, and lasts about 29.5 days.

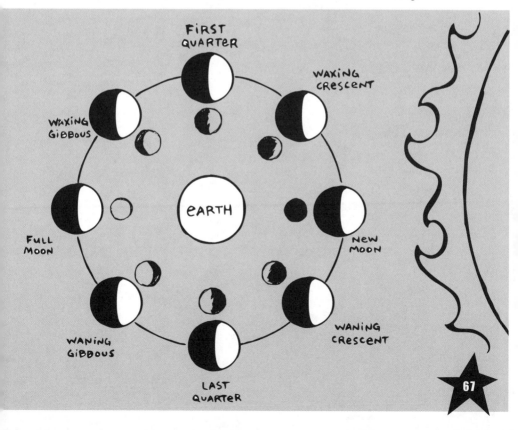

The Harvest Moon

In the Northern Hemisphere, the full moon nearest the

Autumnal equinox is known as the **Harvest Moon** and holds this distinction because of its curiously large and bright appearance. During this time, the Moon's orbit is such that it rises gracefully over the eastern horizon in the early evening, rather than ascending straight towards its zenith. Because of the extra light from this moon, so the tradition goes, farmers can work in the fields, collecting the autumn harvest, well past the setting of the Sun. The effect is similar at the time of the subsequent full moon - the Hunter's Moon - near the end of October.

To the human eye, the Moon in general appears larger over the horizon than it does higher in the sky - a phenomenon known as the *moon illusion*. No one knows for certain why this is so. Several theories exist, but the standard one is attributed to **Ptolemy**, the second century Alexandrian astronomer, who believed that the moon illusion could be explained by the fact that, near the horizon, we have other objects of reference with which to compare the size of the Moon, which we don't have higher in the sky.

SOLAR AND LUNAR ECLIPSES

If the Moon's orbit around the Earth lay precisely on the plane of the ecliptic, we would see two full eclipses during each lunar month. At each new moon, we would see a **total solar eclipse**. The Moon would pass directly between the Earth and the Sun, temporarily casting its shadow onto some parts of the Earth.

OOOH!

AAH!

At each full moon, we would see a **total lunar eclipse**. The Earth would pass directly between the Sun and the Moon, casting its shadow onto the surface of the Moon.

MOON

EARTH

MOON

ECLIPTIC

Of course, this is not the case. In relation to the plane of the ecliptic, the Moon's orbit tilts at an angle of about 5 degrees, and thus the Moon itself varies from as much as 20,000 miles above and below the ecliptic. Furthermore, the two points in the Moon's orbit that intersect the plane of the ecliptic, called the **nodes**, do not remain in a fixed position, but continually revolve around the Earth, as the Earth revolves around the Sun.

In order for an eclipse to take place, the Earth, the Moon, the Sun - and the nodes of the Moon's orbit - must all be in close alignment. Because of the complex cycles of eclipses, there can be no more than five solar eclipses, and no more than three lunar eclipses, in any given year.

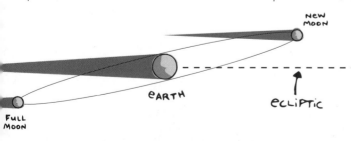

NEW MOON

EARTH

ECLIPTIC

FULL MOON

The Dragon that Devoured the Moon

Before humankind was able to piece together a more accurate understanding of the heavens, solar and lunar eclipses were viewed as highly unnatural events, and regarded with fear and superstition.

In myths and folklore from otherwise diverse cultures throughout the world, these temporary moments of darkness over the Sun or the Moon were interpreted, in one way or another, as evil omens brought on by some sort of malevolent, mythological creature out to devour the venerated heavenly orb. According to various customs, in order to drive away this darkly intrusive force, people would gather outdoors at such times, shouting and singing and banging on drums, until the Sun or the Moon would return safely from the throes of destruction.

AND, APPARENTLY, THIS WORKED EVERY SINGLE TIME!

A solar eclipse can be *total*, when the Moon completely covers the disk of the Sun.

Or it may be a **partial solar eclipse**, when the Moon only partially grazes the disk of the Sun.

In addition, there is also something called an **annular solar eclipse**. We have already seen how the Moon varies in its orbit between apogee and perigee. By a fortunate coincidence, its average size as seen from the Earth is just about the same size as the disk of the Sun as seen from Earth. However, when the Moon is closer to apogee, it is not large enough to completely cover the disk of the Sun, so that when an annular solar eclipse reaches its maximum, a large ring of the Sun is seen around the Moon!

A lunar eclipse can also be total, when the Earth's shadow completely covers the face of the Moon.

Or it may be a **partial lunar eclipse**, when the Earth's shadow only partially covers the Moon.

71

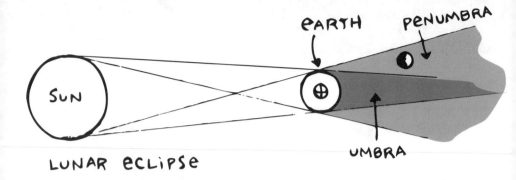

eARTH   PeNUMBRA

SuN

UMBRA

LUNAR eCLiPSe

Furthermore, there is also the phenomenon of a **penumbral lunar eclipse**. The Earth and the Moon cast two different types of shadows in the light of the Sun. The **umbra** is the darker, more narrowly focused shadow, and the **penumbra** is the lighter, more broadly focused shadow.

During a penumbral lunar eclipse, it is only the vague, penumbral shadow of the Earth that passes across the face of the Moon.

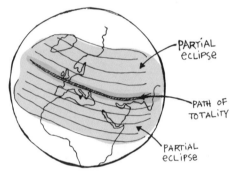

PARTiAL eCLiPSe

PATH OF TOTALiTY

PARTiAL eCLiPSe

During a solar eclipse, the Moon's smaller and more focused umbral shadow hits the Earth. Because of this - and because the Moon itself is so much smaller than the Earth - a solar eclipse of any kind is *best* seen when observed from locations within the narrow path of the Moon's umbral shadow.

Of course, if you have the opportunity to observe a solar eclipse, you should never look at it direct-ly, as it could burn your retinas to the point of blindness! The only safe ways to observe a solar eclipse are either to project the image with a pinhole projector, or to view it through a strong solar filter. (The best thing to do is to get in touch with your local planetarium or observatory to learn more about what they may have to offer.)

WHILE A SOLAR ECLIPSE CAN BE DIFFICULT TO SEE, A LUNAR ECLIPSE IS JUST THE OPPOSITE. BECAUSE THE MOON ACTS AS A KIND OF GIANT PROJECTION SCREEN, A LUNAR ECLIPSE OF ANY KIND CAN BE SEEN EQUALLY WELL FROM JUST ABOUT ANYWHERE ON THE NIGHT SIDE OF THE EARTH FACING THE MOON.

The Eclipse that Saved the Day!

The Greek historian, **Herodotus**, writing in the fifth century BC, tells of a war in Asia Minor between the Lydians and the Medes, which had lasted for five years, with neither side appearing to be winning. In the sixth year, however, in the midst of a fierce battle, the Moon suddenly eclipsed the Sun, and darkness fell over the battlefield. As luck would have it, both sides interpreted this as a sign from the gods that they should immediately make peace - and so they did! Because astronomers are able to calculate the cycles of eclipses both forward and backwards in time, we now know that this event must have occurred on May 28th, in the year 585 BC.

CHAPTER 4:

THE NIGHT SKY

Our Universe, contains at least a billion trillion stars, but on the clearest of nights we're lucky if we can see about 2,000 of them. The closest stars to our own are the three stars of the Alpha Centauri star system, which are about 4.3 light years away.

A LIGHT YEAR is the distance that light can travel in one year.

NOTHING TRAVELS FASTER THAN THE SPEED OF LIGHT,

which cruises along at 186,000 miles per second (300,000 km/second). In one year, light can cover nearly 6 trillion miles (9.5 trillion km), so even the closest stars are almost unimaginably far away.

The **celestial poles** are imaginary points in the sky, projected outwards from the Earth's poles. Because the Earth's axis is consistently tilted in the same direction, at 23.5 degrees in relation to the plane of the ecliptic, the North Pole is always inclined towards the North Star, **Polaris**, even during the daytime, and at all points along the Earth's 600 million mile orbit around the Sun. This is because Polaris is about 400 light years - very,

very

far - away!

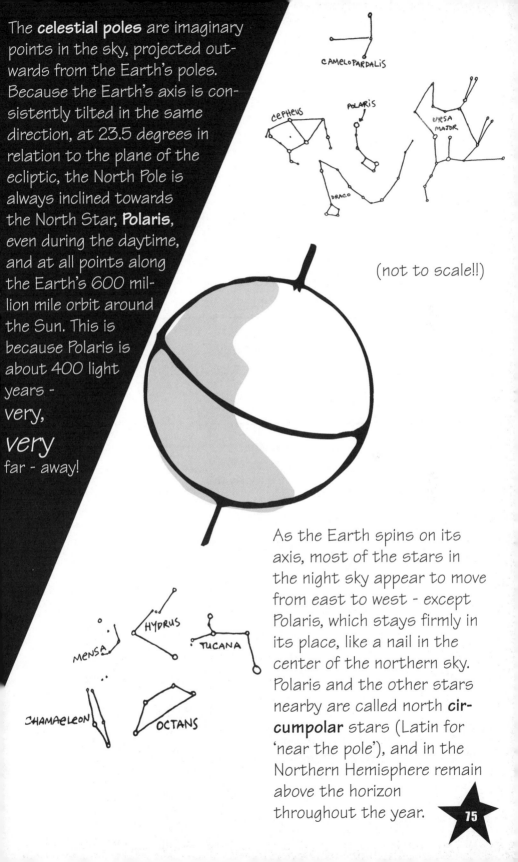

CAMELOPARDALIS

CEPHEUS

POLARIS

URSA MAJOR

DRACO

(not to scale!!)

HYDRUS

MENSA

TUCANA

CHAMAELEON

OCTANS

As the Earth spins on its axis, most of the stars in the night sky appear to move from east to west - except Polaris, which stays firmly in its place, like a nail in the center of the northern sky. Polaris and the other stars nearby are called north **circumpolar** stars (Latin for 'near the pole'), and in the Northern Hemisphere remain above the horizon throughout the year.

The easiest way to locate the North Star is to find the Big Dipper, which is located in the bright, north circumpolar constellation of Ursa Major (Latin for 'Great Bear'), and trace an imaginary line connecting the

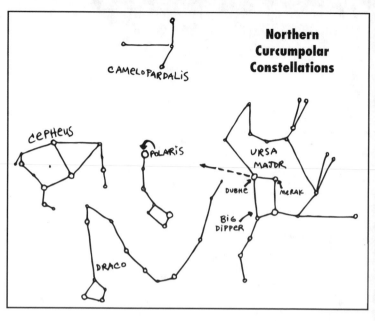

stars Merak and Dubhe about six times their distance from each other. They point almost directly towards Polaris, which is also the tip of the Little Dipper's handle.

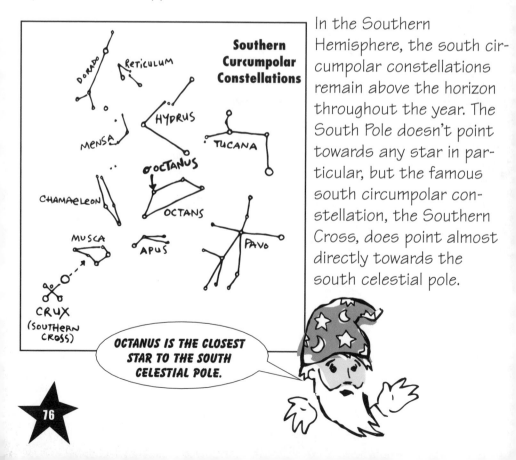

In the Southern Hemisphere, the south circumpolar constellations remain above the horizon throughout the year. The South Pole doesn't point towards any star in particular, but the famous south circumpolar constellation, the Southern Cross, does point almost directly towards the south celestial pole.

OCTANUS IS THE CLOSEST STAR TO THE SOUTH CELESTIAL POLE.

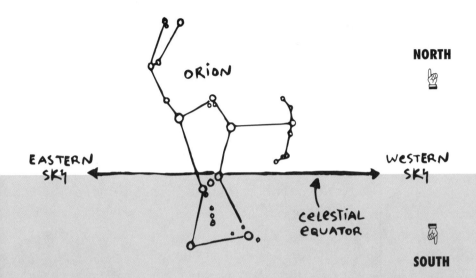

Non-circumpolar stars - stars that are not near the poles - are found lower in the sky, and are visible at different times throughout the year, relative to our orbit around the Sun. For example, in January, the non-circumpolar constellation of Orion is prominent in the night sky, but six months later, when the Earth is on the opposite side of the Sun, Orion is no longer visible.

The exact locations of all non-circumpolar stars depend on the time of the year, as well as one's latitude. The **celestial meridian** is an imaginary line in the sky running north and south, directly above the observer. The **celestial equator** is an imaginary line in the sky running east and west, projected outward from the Earth's equator. A stargazer standing at the Earth's equator would thus see the stars of the celestial equator directly overhead, running east and west. North of the equator, these same stars would be seen lower in the southern sky, while south of the equator, the same stars would be seen lower in the northern sky.

THE BEST WAY TO LOCATE THE STARS IN <u>YOUR</u> NIGHT SKY IS THROUGH THE USE OF A STAR CHART MADE FOR YOUR GENERAL LATITUDE.

The Myth of the Great Bear

Ursa Major is an extremely prominent constellation in the northern sky, and has captivated our attention since ancient times. Today we primarily think of it as the Big Dipper, and Homer tells us that the ancient Greeks also used to think of it as the Wagon.

Its name, however, means 'Great Bear', and this is its story...

Jupiter, the king of the gods, was up to his usual, amorous self, when he fell in love with the woman Callisto, whom he, in due course, impregnated.

On discovering this, Jupiter's jealous wife, Juno, became characteristically furious, and as punishment, after Callisto had given birth to her son, Juno had Callisto transformed into a bear.

But as if this were not vengeance enough, when Callisto's son, Arco, grew older, Juno tried to connive a situation in which the young lad, while out hunting, would end up killing his own mother. Completely unaware of what he was actually doing, he sighted the bear, pulled back his arrow, and released it from its string.

Fortunately, just at that moment, Jupiter intervened, rescuing Callisto and placing her in the heavens in the form of the Great Bear. Later, her son would also come to join her in this privileged place, in the form of Ursa Minor, or the Lesser Bear (also known to us today as the Little Dipper).

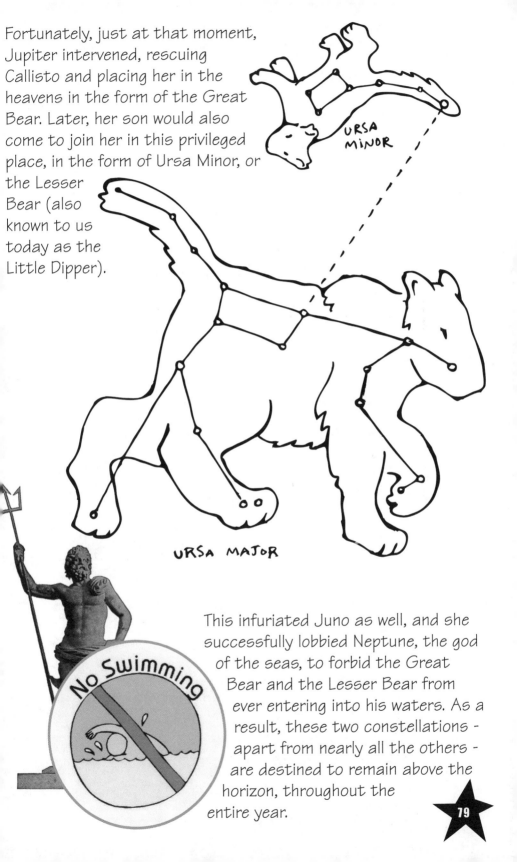

URSA MINOR

URSA MAJOR

This infuriated Juno as well, and she successfully lobbied Neptune, the god of the seas, to forbid the Great Bear and the Lesser Bear from ever entering into his waters. As a result, these two constellations - apart from nearly all the others - are destined to remain above the horizon, throughout the entire year.

No Swimming

The Zodiac

The twelve constellations of the zodiac (from a Greek word meaning 'little animals') are non-circumpolar constellations, which lie along the path of the ecliptic - the same general path taken across the sky by the Sun, the Moon, and the planets. According to ancient superstition, a person's astrological sign is determined by the constellation through which the Sun passes on the day of his or her birth. (That same constellation would not be visible in the *night* sky until about six months later, when the Earth is on the opposite side of the Sun.)

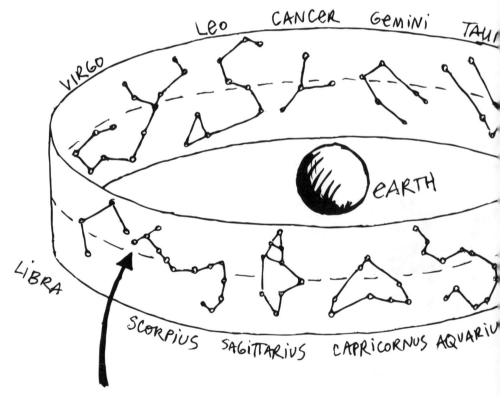

THe SuN'S APPAReNT PATH THROuGH THe CoNSTeLLATIONS

However, aside from there being absolutely no logical reason why the zodiacal constellations should have any effect on our personality or behavior, another difficulty with astrology is something called the **precession of the equinoxes**. This refers to the fact that the gravitational tugs of the Sun and the Moon cause the tilt of the Earth's axis to shift over time. Because of the precession of the equinoxes, the celestial poles slowly change their positions with respect to the stars. In about 13,000 years from now, the North Pole will point towards a new North Star: Vega. But in another 13,000 years, it will again point towards Polaris.

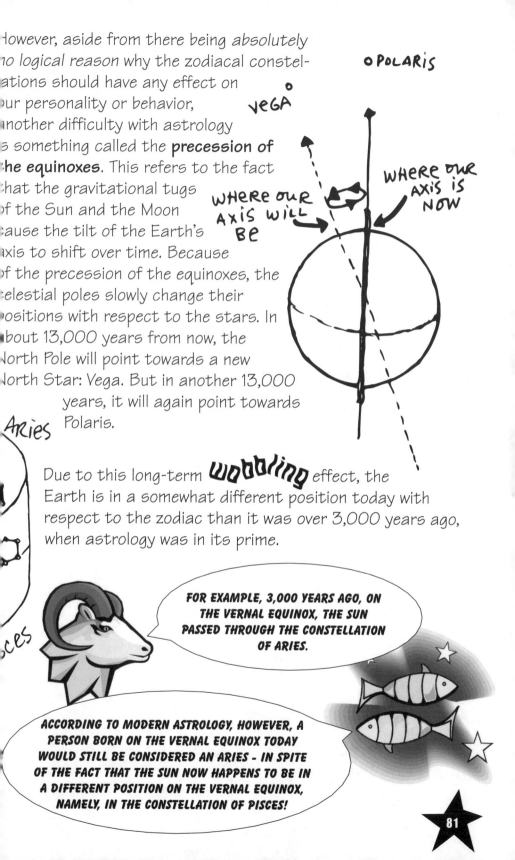

O POLARIS

VEGA

WHERE OUR AXIS WILL BE

WHERE OUR AXIS IS NOW

Aries

Due to this long-term **wobbling** effect, the Earth is in a somewhat different position today with respect to the zodiac than it was over 3,000 years ago, when astrology was in its prime.

FOR EXAMPLE, 3,000 YEARS AGO, ON THE VERNAL EQUINOX, THE SUN PASSED THROUGH THE CONSTELLATION OF ARIES.

ACCORDING TO MODERN ASTROLOGY, HOWEVER, A PERSON BORN ON THE VERNAL EQUINOX TODAY WOULD STILL BE CONSIDERED AN ARIES - IN SPITE OF THE FACT THAT THE SUN NOW HAPPENS TO BE IN A DIFFERENT POSITION ON THE VERNAL EQUINOX, NAMELY, IN THE CONSTELLATION OF PISCES!

THE CELESTIAL SPHERE

The **celestial sphere** is an imaginary sphere used by astronomers to depict the stars as seen from the Earth. With the Earth at its center, the Earth's poles point directly towards the celestial poles, the Earth's equator extends outwards to describe the celestial equator, while the path of the ecliptic crosses the celestial equator at an angle of about 23.5 degrees.

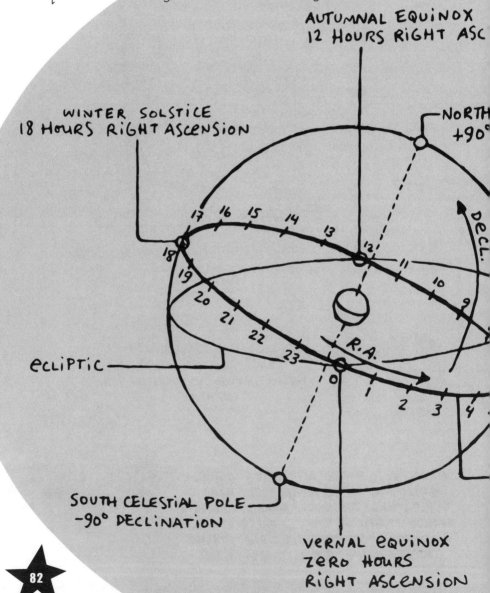

AUTUMNAL EQUINOX
12 HOURS RIGHT ASC

WINTER SOLSTICE
18 HOURS RIGHT ASCENSION

NORTH
+90°

DECL.

ecliptic

R.A.

SOUTH CELESTIAL POLE
-90° DECLINATION

VERNAL EQUINOX
ZERO HOURS
RIGHT ASCENSION

The celestial sphere is used as a map to locate objects in the sky with celestial coordinates. Just as the Earth's equator designates zero degrees latitude, the celestial equator designates zero degrees **declination**. North of the celestial equator, declination is measured in positive numerical degrees (from zero to +90). South of the celestial equator, declination is measured in negative numerical degrees (from zero to -90). Thus, the north celestial pole lies at +90 degrees declination, and the south celestial pole lies at -90 degrees declination.

Whereas the celestial latitude of declination is measured in degrees, minutes, and seconds, the celestial longitude of **right ascension** is measured in *hours*, minutes, and seconds. Right ascension begins at the point on the celestial sphere that depicts the stars as seen on the celestial meridian at the time of the vernal equinox. The celestial sphere is then divided into 24 vertical hour lines: The line of the vernal equinox is the zero hour line, the line of the summer solstice is the 6 hour line, the autumnal equinox is the 12 hour line, and the winter solstice is the 18 hour line.

A star lying exactly on the zero hour line *and* on the celestial equator would thus have the celestial coordinates: zero hours right ascension, zero degrees declination, while a star lying on the 12 hour line, 10 degrees above the celestial equator, would have the coordinates: 12 hours right ascension, 10 degrees declination.

AL POLE
ATION

——SUMMER SOLSTICE
6 HOURS RIGHT
ASCENSION

STIAL EQUATOR
DECLINATION

TYPES OF STARS

The universe is populated by many different types of stars, which vary on the scales of age, size, density, pressure, and brightness. At their cores, stars are capable of a number of different types of nuclear reactions, which correspond to different stages of their natural lifetimes.

Stars are primarily composed of hydrogen and helium, but different types of nuclear reactions can produce different types of atoms, or chemical elements. When different types of elements are burned, they give off different colors, which can be seen in the details of the spectrum of light emitted by a star. (And the science that studies the spectra of light is called spectroscopy.)

By analyzing the color of a star's light, scientists are therefore able to discover information about its temperature and chemical composition, which, in addition to its size, can also indicate its stage of evolution.

As a result, it is known that stars that give off a bluish-white light are generally the youngest and the hottest stars.

Stars that are yellow seem to be in their prime.

Red and orange stars are older and cooler, and

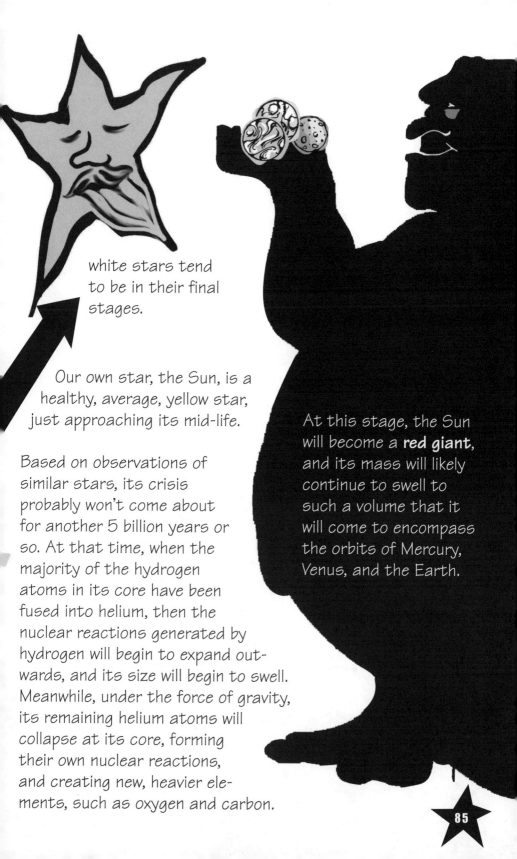

white stars tend
to be in their final
stages.

Our own star, the Sun, is a
healthy, average, yellow star,
just approaching its mid-life.

Based on observations of
similar stars, its crisis
probably won't come about
for another 5 billion years or
so. At that time, when the
majority of the hydrogen
atoms in its core have been
fused into helium, then the
nuclear reactions generated by
hydrogen will begin to expand outwards, and its size will begin to swell.
Meanwhile, under the force of gravity,
its remaining helium atoms will
collapse at its core, forming
their own nuclear reactions,
and creating new, heavier elements, such as oxygen and carbon.

At this stage, the Sun
will become a **red giant**,
and its mass will likely
continue to swell to
such a volume that it
will come to encompass
the orbits of Mercury,
Venus, and the Earth.

As the Sun's volume expands, its temperature will cool. Perhaps as soon as a half a billion years later, when it has burned off the bulk of its matter and can no longer generate the temperatures necessary for nuclear reactions, its remaining elements will steadily shrink, collapsing upon their own gravity to become a small **white dwarf** - about as small as the Earth, but incredibly dense. At this stage, although the Sun will no longer be capable of nuclear reactions, it will nevertheless continue to shine from the heat generated by its own gravitation;

Our Sun

that is, until it ultimately dies off completely, perhaps to become a **black dwarf** - the dark and lifeless chemical remains

A star much larger than the Sun, such as a **red** or a **blue super giant**, will undergo a much more dramatic fate when it expires, becoming a **supernova** - exploding brilliantly under the weight of its gravitational collapse. The bulk of the star's matter and energy is then released back into the universe in the form of an interstellar nebula.

However, at the core of the ex-star, bare neutrons will often remain as by-products of these massive atomic explosions. Bonded together by nuclear forces, these neutrons can then condense to form a **neutron star**, which can shrink down to as little as just 10 miles (16 km) in diameter, but becoming so dense that a teaspoonful of its matter would weigh about a billion tons on Earth!

THAT'S HEAVY MAN!!!

A neutron star can also become a **pulsar** - a type of star that continuously emits pulses of radio waves out into space upon each rotation, which can occur within just fractions of a second!

The very largest stars will also explode in supernovae, but rather than turning into neutron stars, such stars can turn into *black holes* - extraordinary phenomena so dense that nothing can escape their gravitational pull - not even light itself. Naturally, a black hole is rather difficult to see, but astronomers have been able to detect black holes by observing the gravitational effects on objects around them

Another type of star, called a **variable star**, will display a fluctuating brightness, which can have a number of causes.

A star's variability may be *intrinsic*, when its brightness is actually changing. Examples of intrinsic variables are eruptive stars undergoing explosions, and pulsating stars, which vary in brightness due to imbalances between their cores and their outer layers.

A star's variability may also be *extrinsic*, or caused by external factors, such as when one star is periodically eclipsed by another. Variable stars called **Cepheids** will display their variations at a quick and steady pace, over days or weeks, while long-term variables can take periods of over a year.

HEY, GET OUT OF MY LIGHT!!!

MAGNIFICENCE!

A star's brightness is measured in terms of its magnitude.

Absolute magnitude measures the intrinsic brightness of stars, based on how they would be seen from a *uniform* distance.

Apparent magnitude measures the brightness of stars as seen from the Earth, determined by both their absolute magnitude and by their *actual* distance from the Earth.

The brighter the star is, the lower the number of its magnitude. First magnitude stars, with a magnitude of 1.5 or less, are 100 times brighter than sixth magnitude stars, with magnitudes between 5.5 and 6.5.

JUST BECAUSE YOUR MAGNITUDE IS HIGHER DOESN'T MEAN YOU'RE BRIGHTER!!

The brightness of the planets is also judged by magnitude.

5.6 Uranus, which is barely visible to the naked eye, has an apparent magnitude of about 5.6,

-4

while Venus, the brightest planet, has an apparent magnitude of -4.

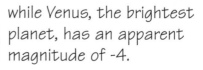

As for the brightest stars in our skies... with an apparent magnitude of 0.3, **Vega** is the fifth brightest star, found in the northern, non-circumpolar constellation of Lyra, which is prominent in the late summer and early autumn. Vega is a young blue star, about 26 light years away, which is orbited by its own solar system.

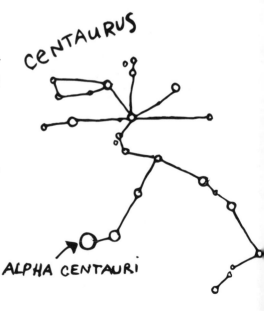

The red giant **Arcturus** is the fourth brightest star in the heavens, with an apparent magnitude of -0.06.
Arcturus is found in the northern, non-circumpolar constellation of Boötes, which is prominent in late spring and early summer.

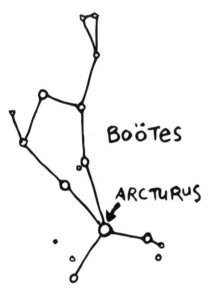

Alpha Centauri is the third brightest star, with an apparent magnitude of -0.1, located in the constellation of Centaurus. Alpha Centauri is actually a triple star: Alpha Centauri A, Alpha Centauri B, and Proxima Centauri C. The three stars orbit each other and, in fact, over half of the stars that we see from the Earth are double, or multiple, stars of this type. To the naked eye, they appear as one, but with a good telescope they can often be distinguished.

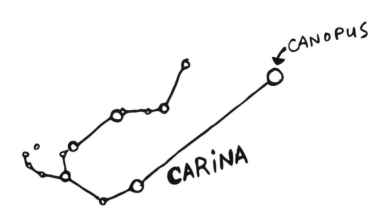

CANOPUS

CARINA

Canopus is the second brightest star, a blue star with an apparent magnitude of -0.7, located in the constellation of Carina.

Centaurus and Carina are both to be found among the south circumpolar constellations.

And the very brightest star as seen from the Earth is the double star of **Sirius**, with an apparent magnitude of about -1.5. Sirius is a blue star orbited by a white dwarf. It is also known as the Dog Star, appearing, as it does, in the constellation of Canis Major (Latin for 'Great Dog'), which is prominent in the winter months, just south of the celestial equator. Six months later, in the hottest days of the northern summer, Sirius and the Sun rise together, and this is where we get the expression, 'the dog days of summer'!

SIRIUS

CANIS MAJOR

STAR CLUSTERS AND NEBULAE

When two stars orbit each other, like the double star system of Sirius, they are also referred to as **binary star systems**, while stars systems such as Alpha Centauri are called **triple star systems**.

Star systems can contain practically any number of stars, and the very largest systems, called **star clusters**, can contain hundreds and even millions of stars bound together by gravity.

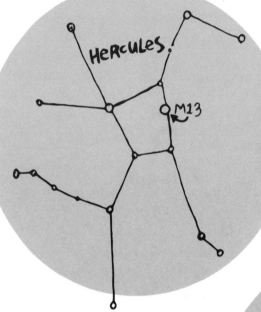

There are two main types of star clusters. **Globular clusters** are immense collections of stars that are spherical in shape. One great example of a globular cluster is the Great Globular Cluster (also called M13), which contains over a half a million stars and can be seen in the northern, non-circumpolar constellation of Hercules during the summer months.

In the Southern Hemisphere, the globular cluster of Omega Centauri can be found in the south circumpolar constellation of Centaurus. Omega Centauri contains several million stars, and is the largest and brightest globular cluster in the galaxy.

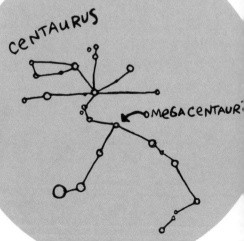

Open clusters are usually smaller clusters with irregular shapes.

For example, the star system of the Pleiades is an open cluster containing around 500 stars. To the naked eye, the Pleiades appear to be just seven stars, but with a pair of binoculars or a telescope many more can be made out.

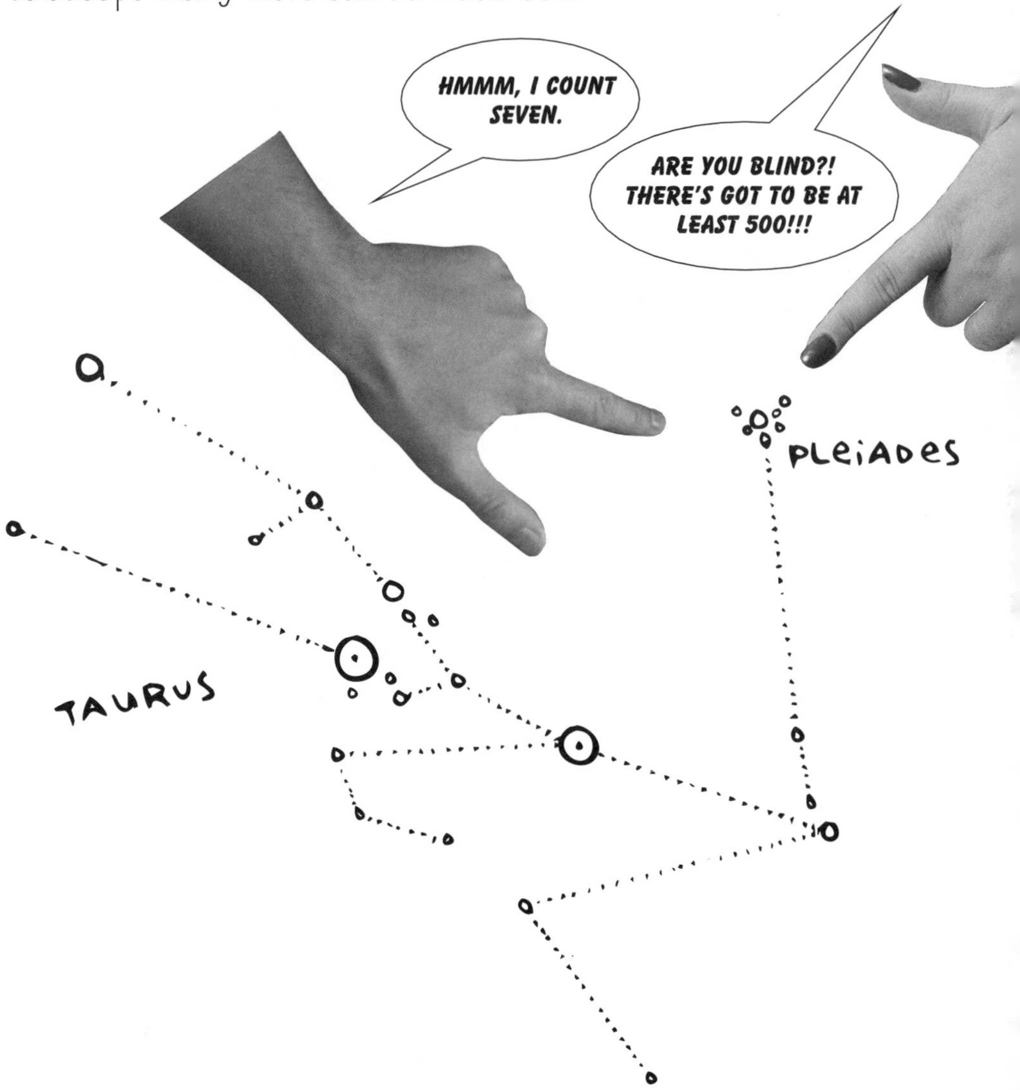

HMMM, I COUNT SEVEN.

ARE YOU BLIND?! THERE'S GOT TO BE AT LEAST 500!!!

PLEIADES

TAURUS

The Pleiades can be found in the zodiacal constellation of Taurus, which is prominent in the winter months.

The sky also contains several different types of nebulae.

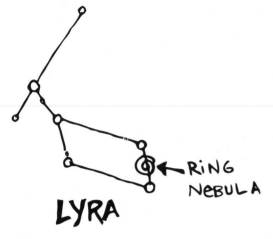

RING NEBULA

LYRA

A nebula that forms as the result of an exploding star is called a **planetary nebula**. A good example of this type of nebula is the Ring Nebula in the northern, non-circumpolar constellation of Lyra.

Emission nebulae are collections of gas and dust that emit light by reacting to the radiation from nearby stars. Such nebulae are very often the birthplaces of new stars. The Great Nebula (also called M42) is an emission nebula in the constellation of Orion, which lies on the celestial equator and is prominent in the winter months.

Among the stars of the Pleiades are also a number of **reflective nebulae,** which reflect the light from nearby stars. By contrast, **dark nebulae** are collections of gas and dust that block or diminish the light from nearby stars, such as the Horsehead Nebula in Orion.

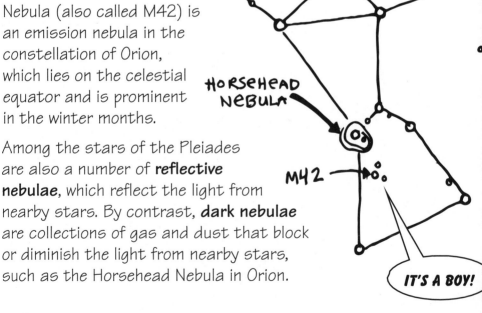

ORION

HORSEHEAD NEBULA

M42

IT'S A BOY!

The Pleiades and Orion

According to Greek and Roman myth, the
seven sisters of the Pleiades were pursued by the hunter, Orion, for seven inter-
minable years until the gods finally
answered their prayers for deliverance,
transforming them into doves and
placing them in the heavens. But as fate
would have it, upon his death, Orion,
too, was placed in the heavens,
so that even today he
continues to pursue them
across the night sky!

COMETS & METEORS

Our solar system is home not only to the nine planets, their moons, the asteroid belt, and the Kuiper belt, but also to a large number of **comets** (Latin for 'hairy star') - frozen bodies of gas and dust that periodically zoom in, whirl around the Sun, and zoom out!

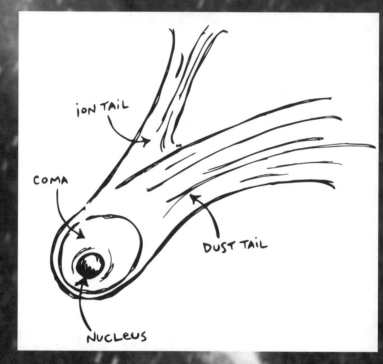

At its core, a comet is composed of an icy, rocky head, or **nucleus**. As it approaches the Sun, however, solar radiation begins to melt and vaporize the comet, forming a cloudy halo, or **coma**, around the nucleus. Solar winds blowing at the comet also create the comet's **tail**, which can sometimes reach millions of miles in length!

THE SUN IS GIVING ME A COMA!!!

A comet's orbit is highly elliptical. Short-term comets orbit the Sun within short periods - around 200 years or less - while long-term comets can take thousands of years or even longer to return to the Sun.

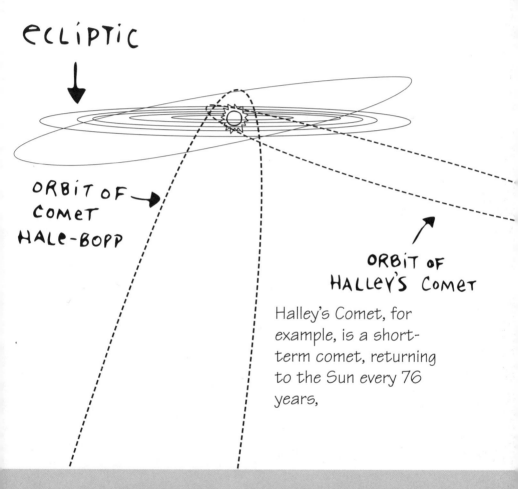

ecliptic

ORBIT OF COMET HALe-BOPP

ORBIT OF HALLeY'S COMeT

Halley's Comet, for example, is a short-term comet, returning to the Sun every 76 years,

while Comet Hale-Bopp, which graced our skies in 1996 and 1997, was an example of a long-term comet, visiting the Sun only once about every 4,000 years.

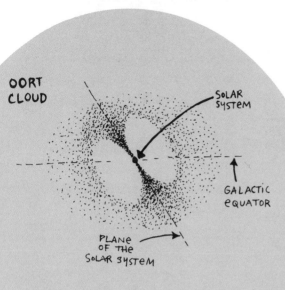

OORT CLOUD

SOLAR SYSTEM

GALACTIC EQUATOR

PLANE OF THE SOLAR SYSTEM

Many short-term comets are believed to originate in the Kuiper belt, perhaps caused by collisions between large Kuiper objects, knocking some of them off of their regular orbits and into the direction of the Sun. The origin of long-term comets is not known for certain, but one possibility is the existence of an immense cloud of cometary material, called the **Oort Cloud**, surrounding the solar system at a distance of perhaps 50,000 AU from the Sun.

As comets whirl around the Sun, they leave a trail of debris in their wake, known as **meteoroids**. Some of these streams of meteoroids develop orbits around the Sun that intersect with our own, and so at certain points throughout the year we run right into them.

When a meteoroid enters the Earth's atmosphere, it becomes a **meteor**, or a 'shooting star', burning up from friction as it falls from the sky.

If it manages to reach the Earth's surface, it is designated a **meteorite**.

While shooting stars can occasionally be seen on just about any clear, dark night of the year, the *best* times to see them are during one of the year's major meteor showers.

FOR EXAMPLE:

Around the night of January 3rd, the Earth runs right into the **Quadrantids,** a meteor shower so named because it was once associated with an archaic constellation known as Quadrans Muralis. These days, the Quadrantids can best be seen coming from an area in the northern, non-circum-polar sky where the con-stellations of Hercules, Boötes, and Draco come together.

DRACO

THis is wHeRe you caN see THe QUADRANTiDs

HeRCULeS

BoöTes

For around five days surrounding August 12th, the Earth runs right into the **Perseids,** so named because they can be seen coming from the northern, non-circumpolar constellation of Perseus. The Perseids were last left behind in the wake of Comet Swift-Tuttle, whose most recent perihelion (closest) passage to the Sun occurred in the summer of 1992.

THe PeRSeiDs Go HeRe

PeRSeUS

99

Around October 21st, our orbit brings us into contact with the **Orionids**, a meteor shower so named because it can be seen coming from the constellation of Orion, which lies on the celestial equator. The Orionid meteoroid stream was last left behind in the wake of Halley's Comet, whose most recent perihelion passage occurred in February of 1986.

THE ORIONIDS HAPPEN HERE →

ORION

LEO

WHERE THE LEONIDS ARE

Around November 17th, our orbit around the Sun brings us into contact with the **Leonids**, a meteor shower which can be seen coming from the zodiacal constellation of Leo. The Leonid meteoroid stream was last left behind in the wake of Comet Temple-Tuttle, whose most recent perihelion passage occurred in February of 1998. (And because of 1998's passage, 1999's Leonid meteor shower was a veritable meteor storm - showering the skies at certain locations with thousands of shooting stars!)

WHERE you see the GEMINIDS

For around three days surrounding December 14th, our orbit around the Sun brings us into contact with the **Geminids**, a meteor shower so named because it can be seen coming from the zodiacal constellation of Gemini. The Geminid meteoroid stream may not actually be related to any comet, but rather may be the debris from an asteroid called Phaeton, whose eccentric orbit coincides perfectly with the Geminids.

GEMINI

The debris that enters our atmosphere to become a meteor is usually quite small - often just the size of a grain of dust! On rare occasions, however, a meteor can be so large that it can become a **fireball**, burning brilliantly as it falls from the sky. Sometimes a fireball can be so bright that it can actually be seen in the daytime sky.

Comets as Prophets

As with solar and lunar eclipses, many cultures throughout the ages have widely regarded comets and meteors as evil omens, often thought to fore-shadow such calamities as pestilence, war, or the death of a king.

According to one famous story, when Halley's Comet appeared in 1456, in the middle of the Crusades, the pope at the time, Callixtus III, went so far as to have it excommunicated!

A few exceptions to this gloomy outlook have existed, however. For example, the !Kung people of Namibia took the highly unusual position that comets were omens of good things to come.

Similarly, a story is told that when a comet appeared shortly after the death of Julius Caesar, the new Roman emperor, Augustus, declared it as a sign that Caesar had joined the gods in heaven and went on to have the comet venerated in a Roman temple.

HEY LOOK! IT'S CAESAR!!

GALAXIES

All of the stars in the night sky that are visible to the naked eye belong to our own galaxy, the Milky Way, although other distant galaxies are also visible. The **Magellanic Clouds**, for example, are smaller galaxies orbiting the Milky Way, which can be seen in the southern circumpolar sky.

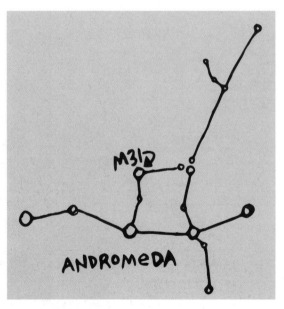

The **Andromeda Galaxy** (also called M31) is the most distant object in the night sky that can be seen with the naked eye. It appears as a fuzzy patch of light in the northern, non-circumpolar constellation of Andromeda, and is prominent in the autumn months.

Surrounded by many smaller galaxies, the Milky Way and the Andromeda Galaxy are the two largest galaxies in our **Local Group**, a group of more than twenty nearby galaxies (relatively speaking!).

Groups of galaxies are furthermore combined into **galaxy clusters**, which can contain hundreds and thousands of galaxies, all attracted to each other by the force of gravitation and moving together through space. Our own Local Group belongs to the massive **Virgo Cluster**, which may contain tens of thousands of galaxies.

Galaxies can come in all shapes and sizes.

Elliptical galaxies are spherical in shape, with bright centers, and are composed primarily of older stars.

Spiral galaxies have a flattened, disk-like shape when viewed from the side, but show circular, spiral patterns when viewed from above. Spiral galaxies contain both old and new stars, as well as molecular clouds and nebulae.

Some spiral galaxies, called **barred spirals**, display streams of gaseous material, shooting out from their centers, and often connecting with their spiral arms.

Other galaxies that are flattened and disk-shaped, but lacking in spiral patterns, are called **irregular galaxies** and are primarily composed of young stars and nebulae.

eLLiPTiCAL

SPiRAL

BARReD SPiRAL

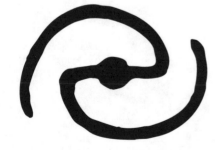

The Milky Way and the Andromeda Galaxy are both spiral, disk-shaped galaxies, containing hundreds of billions of stars rotating around a common center.

The center of the Milky Way Galaxy, around which we rotate about every 250 million years, lies about 26,000 light years away. It can be seen as a hazy band of light on clear, moonless nights, centered in the zodiacal constellation of Sagittarius, which is prominent in the late summer.

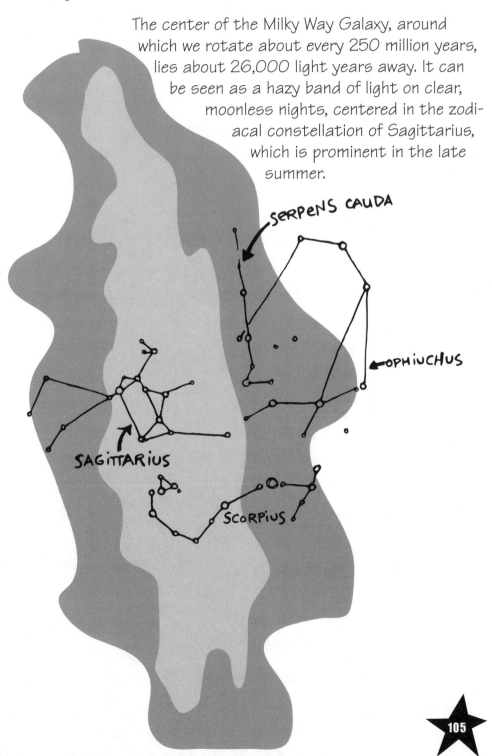

SERPENS CAUDA

OPHIUCHUS

SAGITTARIUS

SCORPIUS

CHAPTER 5:

MEASURING TIME

We have already seen how our orbit around the Sun defines the Earth's seasons and the length of our year, but different astronomical cycles are also responsible for just about every other way we mark time as well.

THE MONTH

The lunar month is the period of time between successive new moons - roughly 29.5 days. Because of the visibility and reliability of the Moon and its phases, for thousands of years, human cultures divided the year into units of lunar months.

(In fact, the words *moon* and *month* both derive from the same Indo-European word, whose linguistic origin is so ancient that it pre-dates written language.)

THE ONLY PROBLEM WAS THAT 12 LUNAR MONTHS OF 29.5 DAYS WERE ABOUT 11 DAYS TOO SHORT FOR ONE 365-DAY SOLAR YEAR,

AND 13 LUNAR MONTHS WERE ABOUT 18 DAYS TOO LONG.

Several attempts at reconciling the two were made - all of which proved to be rather difficult...

The ancient Babylonians were the first to develop a **lunisolar** calendar, that is, a calendar based primarily on lunar months, but with *intercalary* months, or

'leap months',

added every so often to accommodate the solar year.

The ancient Egyptians tried to

extend

the lunar month from 29.5 days to 30 days. Twelve 30-day months yielded a 360-day year, to which they added 5 religious days of celebration. This was fine until they realized the need for another extra leap day every four years - an idea so radical that it wouldn't be adopted for nearly two hundred years after it was first proposed by the king of Egypt, Ptolemy III, in the third century BC.

Eventually, the Romans adopted from the Egyptians the notion of the solar year, but they solved the lunar/solar problem by cutting the Gordion knot! They simply rearranged the lengths of the months, essentially replacing the lunar month with a civil month, a tradition our modern calendar has inherited.

Today, the traditional Jewish and Chinese calendars still retain the lunisolar system, while the Islamic calendar is solely lunar, adhering strictly to lunar months. The solar year is of no account and, because of this, the Islamic months arrive at different times and in different seasons of the solar year.

THIS IS THE HOTTEST DECEMBER WE'VE HAD IN A WHILE!

THE WEEK

NEW MOON

FIRST QUARTER

FULL MOON

LAST QUARTER

Compared to other measurements, the length of the week seems pretty arbitrary. At one point in time, the ancient Assyrians had a six-day week, the Romans had an eight-day week, and the Greeks and the Egyptians had a ten-day week. Some historians trace the origin of the seven-day week to the Babylonians. Others believe that it originated with the Hebrews, stemming from the seven biblical days of creation. It is also quite probable that the Hebrews adopted the seven-day week from the Babylonians during the time of their captivity. In any event, regardless of its exact origin, its use eventually spread throughout the ancient world.

The seven-day week may have found its resonance with the four major phases of the lunar month: the first week beginning with the new moon, the second with the first quarter phase, the third with the full moon, and the fourth with the last quarter.

However, many ancient cultures also believed that the days and the hours were ruled by the Sun, the Moon, and the five known planets, which were synonymous with the gods themselves. Accordingly, the Greeks named the seven days of the week after the Sun, the Moon, and the planets. The Romans and the Norse followed suit, but changed the names according to their own language and mythology, all of which is still reflected in the names we use today.

The Greeks named the first day of the week 'the day of the Sun', *hemera Helio*. The Romans translated this directly into Latin as *dies Solis*. The Anglo-Saxons, in turn, called it *Sunnandaeg*, from which modern English derives the word, Sunday.

The Greeks named the second day of the week 'the day of the Moon', *hemera Selenes*. The Romans directly translated this as *dies Lunae*, and the Anglo-Saxons as *Monandaeg*, from which we derive the word, Monday.

The Greeks named the third day of the week *hemera Areos*, 'the day of Ares', after the Greek god of war. His mythological and planetary counterpart in the Roman pantheon was Mars, and so the Roman called this day 'the day of Mars', *dies Martis*. The Norse version of Mars was the god Tyr, which evolved into the Anglo-Saxon as Tiw, and from the Anglo-Saxon *Tiwsdaeg* we derive the word, Tuesday.

MARS

The Greeks named the fourth day of the week *hemera Heru*, 'the day of Hermes', after the Greek god of travel, commerce, thievery, and cunning. His counterpart in the Roman pantheon was Mercury, and so the Romans called this day 'the day of Mercury', *dies Mercurii*. The Norse counterpart was Woden, and from the Anglo-Saxon *Wodensdaeg* we derive the word, Wednesday.

MERCURY

JUPITER

The Greeks named the fifth day of the week *hemera Dios*, 'the day of Zeus', after their supreme god of thunder and the heavens. His Roman counterpart was Jupiter, and so the Romans called this day 'the day of Jupiter', *dies Jovis*. The Norse counterpart was Thor, and from the Anglo-Saxon *Thorsdaeg* we derive the word, Thursday.

The Greeks named the sixth day of the week *hemera Aphrodites*, 'the day of Aphrodite', after the Greek goddess of love. Her Roman counterpart was Venus, and so the Romans called this day 'the day of Venus', *dies Veneris*. The Norse counterpart was Freya, which evolved into the Anglo-Saxon as Frigg, and from the Anglo-Saxon *Frigesdaeg* we derive the word, Friday.

VENUS

SATURN

Finally, the Greeks named the seventh day of the week *hemera Khronu*, 'the day of Cronos', after the Titan, Cronos, the father of Zeus. His Roman coun-terpart was Saturn, and so the Romans called this day 'the day of Saturn', *dies Saturni*. The Norse chose not to reformulate this day, and so, from the Anglo-Saxon *Saetrdaeg* we derive the word, Saturday.

THE DAY

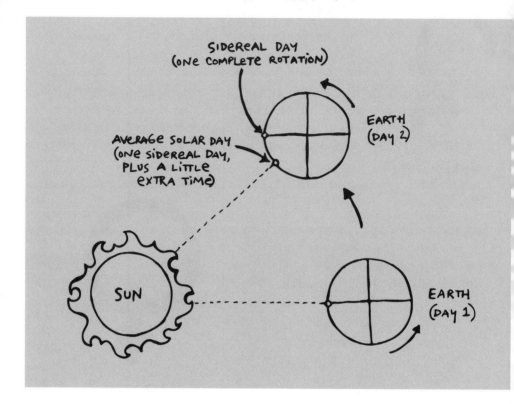

It may come as a surprise that there is actually more than one way to define the day. The **sidereal day** is the time that it takes for the Earth to complete one rotation with respect to the stars - roughly 23 hours and 56 minutes. The sidereal day is different from the **solar day**, which is the time that it takes for the Earth to complete one rotation with respect to the Sun. On average, the solar day is 24 hours long - one sidereal day, plus a little extra time for the Earth to move along in its orbit, in order for the same meridian to 'catch up' with its position to the Sun.

However, because of two factors - the tilt of the Earth's axis of rotation and our elliptical orbit around the Sun - our exact period of rotation with respect to the Sun is not actually consistent throughout the year.

Let's imagine an alternative solar system, in which the Earth's axis of rotation is directly perpendicular to a perfectly circular orbit around the Sun. If this were the case, then the solar day would always be 24 hours long, the day and the night would always be of equal length, and the path of the Sun would consistently follow the path of the celestial equator. Furthermore, the Sun would always cross the same celestial meridian at noon, and sunrise and sunset would occur at the same time each day.

ALTeRNATive SoLAR System:

ReAL SoLAR system:

In the *real* solar system, however, the tilt of the Earth's axis of  rotation causes the Sun's apparent path across the sky to change throughout the year. In the Northern Hemisphere, the path of the Sun is at its highest and northernmost point at the summer solstice, and at its lowest and southernmost point at the winter solstice. In the Southern Hemisphere, just the reverse is true. As a result, from any given point on the Earth, as the length of the Sun's path changes throughout the year, the Sun requires varying amounts of time to cross the sky.

Furthermore, as Kepler's second law informed us, the elliptical path of our orbit changes our distance from the Sun, which also changes our speed. In the winter, when perihelion brings us closer to the Sun, the Earth actually moves a little faster along in its orbit, and in the summer, when aphelion takes us farther from the Sun, the Earth moves a little more slowly along in its orbit.

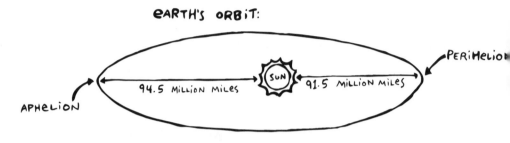

eARTH'S ORBiT:

PERiHELiOM

APHELION

94.5 MiLLiON MiLeS

SUN

91.5 MiLLiON MiLeS

As a result of the combined two effects, the solar days are

regularly
irregular.

Throughout the year, the length of the solar day fluctuates between as much as 16 minutes ahead of 24 hours, to 14 minutes behind 24 hours! Over the course of the year, these variations cancel each other out, so that the *average* length of the solar day is about 24 hours. This is the time we use as the daily world standard, and is called the **mean solar day**.

JUST ANOTHER 'AVERAGED' DAY!

The primary world standard for measuring the mean solar day is called **Universal Time (UT)**, or **Greenwich Mean Time (GMT)** - after the Royal Greenwich Observatory, in Greenwich, England, which lies on the **prime meridian** at zero degrees longitude.

The Greenwich meridian was adopted as the prime meridian in 1884, at the International Meridian Conference, which was called to create a single standard of international time zones, for the purposes of navigation, commerce, and astronomical measurements. As a result, when the fictitious *mean* Sun crosses the prime meridian, it is 12:00 noon UT.

The Earth's 360 degrees of longitude are essentially divided into 24 time zones, and **local standard time** is determined from Universal Time by adding an hour for each time zone to the east of the prime meridian, or subtracting an hour for each time zone to the west.

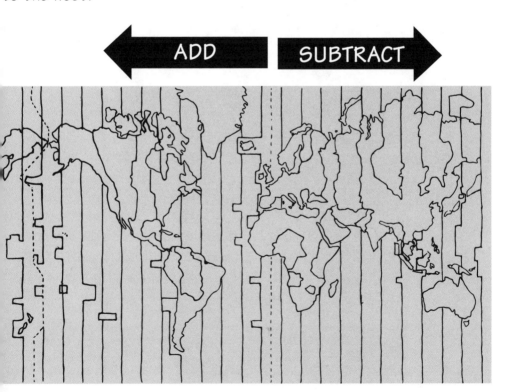

The difference between the true solar day and the mean solar day is referred to as the **Equation of Time**. When the true Sun reaches the prime meridian ten minutes before 12:00 noon UT, the Equation of Time is +10. When the Sun reaches the prime meridian ten minutes after 12:00 noon UT, the Equation of Time is -10.

+20

+15   DIAL FAST (MINS)

+10   JULY   AUG   SEPT   OCT   NOV   DEC

+5   APR   MAY   JUN

0   JAN   FEB   MAR

                    DIAL SLOW (MINS)
-5

-10   On just four days
                                          during the year (April 16th,
-15   June 15th, September 1st, and
                    December 25th), the Equation of Time is zero. On
these days, the true solar day and the mean solar day are both
24 hours long. At 12:00 noon UT, the Sun is directly above the
prime meridian, and at 12:00 noon, local standard time, the Sun
will be directly above the standard meridian of the local time zone

THE ANALEMMA

Because of the difference between true solar time and mean solar time, a sundial will tell time differently than a clock. If you were to mark the shadow of a sundial at the same time each day, over the course of the year, you would trace out an elongated figure-eight pattern known as the **Analemma**. Similarly, time-lapse photography taken of the Sun at the same time each day, over the course of the year, will also show the same figure.

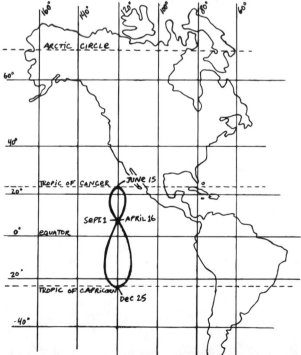

It was once common for the Analemma to be depicted on globes, to represent the Sun's path between the tropics, as well as the Equation of Time.

The height of the Analemma represents the variations of the Sun's height. In latitudes north of the tropic of Cancer, and south of the tropic of Capricorn, the Sun will never be directly above the observer. For an observer north of the tropics, the Sun will always cross the meridian on the south side of the sky, and for an observer south of the tropics, the Sun will always cross the meridian on the north side of the sky. In latitudes between the tropics, the Sun will be directly overhead two times a year.

The width of the Analemma represents the Equation of Time, and the vertical line represents mean solar noon. Where the line and the Analemma meet, the Equation of Time is zero: according to our clocks, it is 12:00 noon UT when the Sun crosses the prime meridian. Conversely, where the Analemma diverges from the line, the Sun is ahead of, or behind, the prime meridian at 12:00 noon UT.

CHAPTER 6: OUR PLACE IN THE COSMOS

One can only imagine the sense of awe in which our earliest ancestors must have lived their lives. When day-to-day survival had to take precedence over intellectual exploration, the Heavens must have seemed the most unfathomable mystery in an utterly mysterious world.

WHAT EXACTLY WAS THE SUN THAT TRAVELED ACROSS THE SKY, BRINGING LIGHT TO THE DAYS, OR THE MOON THAT REPEATEDLY WANED UNTIL IT SEEMED TO DISAPPEAR COMPLETELY, ONLY TO THEN REAPPEAR FROM NOTHINGNESS, WAXING AGAIN UNTIL IT LIT UP THE NIGHT?

WHAT WERE THOSE PINPOINTS OF LIGHT IN THE NIGHT SKY, WHICH SEEMED TO DESCRIBE SUCH MYSTERIOUS SHAPES, OF ANIMALS AND HUNTERS, OR THE PLANETS THAT MOVED AMONGST THEM AS IF BY THEIR OWN ACCORD?

HOW FAR AWAY WERE THESE THINGS?

WERE THEY LIVING CREATURES LIKE US, OR WERE THEY MANIFESTATIONS OF THE GODS THEMSELVES?

There were so many impossible questions, but one thing must have seemed certain above all: The ground upon which we stood was the unmoving center around which the Heavens revolved. This, at least, was an obvious fact that anyone with any sense could see with his own eyes!

In time, the patterns of the heavens and the cycles of the year came to be better understood. As early humans moved from hunting and gathering to an agricultural way of life, astronomical cycles became associated with the agricultural cycles of plowing, sowing, and harvesting the crops. This can be seen in the words of the Greek poet, **Hesiod**, in his poem, *The Works and Days*, written in the ninth century BC, in which he advises his brother to collect the harvest in the fall, and to plow again in the spring:

When the Pleiades, daughters of
Atlas,
are rising, begin your harvest,
and your ploughing
when they are going to set.

[Translator: Hugh G. Evelyn-White (died 1924), Project Guttenberg, www.promo.net/pg/]

An understanding of the night sky also came to be used in celestial navigation upon the seas. The circumpolar stars could be used to determine the cardinal points of north, south, east, and west. And as long as seafarers knew the time of the year, and corresponded the positions of the non-circumpolar stars to their maps of the Earth, they could then determine their general locations as well. This can be seen in *The Odyssey*, in which the Greek poet, **Homer**, sings to us of Odysseus at sea,

He never closed his eyes,
but kept them fixed on the Pleiads,
on late-setting Boötes, and on the Bear
- which men also call the wain,
and which turns round and round where it is,
facing Orion,
and alone never dipping into the stream of
Oceanus.

[Translator: Samuel Butler (1835-1902), Project Guttenberg,
www.promo.net/pg/]

Understanding astronomy thus became a very practical pursuit - but it had a much more mystical side as well. After all, if the Sun controlled the seasons, and if the Moon controlled the tides, then why couldn't the constellations be interpreted as controlling the crops?

It could also follow that the zodiacal constellations, and the planets - which could also be seen as gods - could exert their influence, too, upon the shape and destiny of nations and individuals.

As a result of this type of thinking, for thousands of years, the science of astronomy and the superstition of astrology marched hand in hand.

Various peoples throughout the ancient world came to develop their own systems of mapping the heavens into constellations, using the astronomical cycles to measure the days, weeks, months, seasons, and years - as well as attempting to predict their future with astrology.

But it was the Sumerian and Babylonian civilizations of ancient Mesopotamia (between 4000 and 450 BC) who were the first and most influential progenitors of astronomy and astrology. Carefully observing the positions of the Sun and the Moon, the Mesopotamians were the first to reconcile the lunar month with the solar year, and the first to develop the original signs of the zodiac that are still with us today.

It is also from these civilizations that we have inherited the *sexagesimal* mathematical system, which uses the number 60 as a base, as opposed to the standard *decimal* system, which uses the number 10 as a base. The sexagesimal system divides the circle into 360 degrees, the hour into 60 minutes, and the minute into 60 seconds. It is thus used today in mapping the globe and the celestial sphere, as well as in measuring time itself.

By the sixth century BC, the torch of Mesopotamian astronomy had been passed onto Greece. The Greek philosopher, **Pythagoras** (circa 580 - c.500 BC), and his followers, the Pythagoreans, believed that within the language of mathematics could be found the ultimate answers to the mysteries of the universe. In fact, Pythagoras invented the word, *cosmos*, which he used to refer to the underlying, harmonious order of the universe.

Pythagoras and his followers believed that the forms of the circle and the sphere represented geometrical and mystical perfection, in which, unlike all other shapes, all outside points were equally distant from the center. Because they believed that the cosmos itself was perfect and orderly, they also believed that the Sun, the Moon, and the five known planets were all perfect spheres.

This view was later upheld by the philosopher, **Aristotle** (384 - 322 BC), and the astronomer, **Eudoxus** (408 - 355 BC), who furthermore believed that the sphere of the Earth was the unmoving center of a nice little universe, surrounded and orbited by eight concentric, crystal-clear spheres, to which the stars and the heavenly bodies were attached.

FIRMAMENTUM
SATURNI
IOVIS
MARTIS
SOLIS
VENERIS
MERCVRII
LVNA

Remarkably ahead of his time, **Aristarchus of Samos** (c.310 - c.250 BC) disagreed with this **geocentric**, or Earth-centered model of the universe. By observing the Earth's shadow on the Moon during a lunar eclipse, Aristarchus is said to have determined that the Earth was a planet, which rotated on its axis once every 24 hours, and orbited the Sun with the rest of the planets.

In other words, the solar system was **heliocentric**, that is, it was not centered around the Earth,

but around the Sun.

Aristarchus also determined that the Sun was much larger than the planets, was very far away, and that the rest of the stars were also suns and were even more distant.

Unfortunately, his ideas were rejected at the time, and they wouldn't be revived for nearly 1,800 years.

And so the geocentric model prevailed. But in addition to being incorrect, this model was also flawed in that it failed to account for the observed motions of the planets. While the stars in our sky rotate at a steady and predictable pace, from our point of view on Earth, the planets are wanderers. Generally, from night to night, the planets are seen to travel in the same direction - from west to east against the background of the stars. But every now and then, they *retrogress*, that is, they move backwards. This is referred to as **retrograde motion**. We now know that this is caused by the fact that all of the planets orbit the Sun at different speeds.

In a sense, the solar system is like a giant racetrack, and when the Earth overtakes another planet, that planet's observed motion will be seen to temporarily change. But if the planets orbited an immobile Earth upon perfect spheres, what then could explain their retrograde motion?

In the second century AD, in the Egyptian city of Alexandria, the astronomer, Ptolemy, came up with an answer. In his great work entitled *The Amalgest* (Latin for 'Great Work'), Ptolemy kept most of the details from the old geocentric model: the Earth was the center of the universe, surrounded first by the Moon, then Mercury, Venus, the Sun, Mars, Jupiter, Saturn, and finally, the sphere of fixed stars. But rather than having the heavenly bodies simply orbit the Earth on perfect spheres, he added smaller spheres onto the larger spheres, which he called epicycles. The center of each epicycle was called its deferent. As the planets orbited the Earth, they simultaneously orbited the deferents of their epicycles – which could neatly explain their occasional retrograde motion.

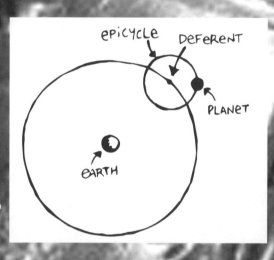

The Ptolemaic system may have been a little complicated. For one thing, it had to give each heavenly body a different rule to follow. But it explained the motions of the heavenly spheres while keeping the Earth at the center of the universe, and so it eventually came to be adopted by the Church during the Middle Ages. As a result, it remained essentially unchallenged for about 1,400 years.

This would be changed, however, by the Polish astronomer and cleric, **Nicolaus Copernicus** (1473 - 1543), who finally brought the heliocentric model back to life, in 1543, with the publication of his masterpiece, *The Revolution of the Heavenly Orbs*. Copernicus believed that the motions of the heavenly bodies could be explained much more simply if the planets orbited the Sun, and if the Earth rotated on its axis. This would also imply that the stars were very far away, and that the universe was much bigger than had previously been supposed. Copernicus still erroneously believed that the Sun was the center of the universe, and that the planets essentially revolved around the Sun in perfectly circular orbits. But in spite of these minor flaws, his contribution was a giant leap forward.

Still, at the time, Copernicus' heliocentric theory was just a theory and, like a lot of new ideas, it encountered a lot of opposition. The Italian astronomer and physicist, **Galileo Galilei** (1564 - 1642), however, was one of the few admirers of the Copernican system. Shortly after the telescope was invented in 1609, Galileo immediately started improving upon it, and he quickly used the new device to observe the heavens, making unprecedented discoveries.

Galileo soon discovered the rings of Saturn, as well as the four largest moons of Jupiter, which were found to orbit Jupiter - not the Earth!

He also studied the Moon itself and discovered that it was not, in fact, a perfect heavenly sphere, but seemed to be covered in craters, mountains, and valleys.

Furthermore, his telescopes also allowed him to be the first person to see multitudes of stars in the sky, beyond those just visible to the naked eye.

He was also able to observe that the planet Venus went through phases like the Moon, indicating that its orbit was centered around the Sun and not the Earth.

All of Galileo's observations seemed to discredit the standing Ptolemaic system, and to support the Copernican system, and he was quick to publish many of his new discoveries in his book, *The Starry Messenger*, in 1610.

But in spite of so much evidence against it, the Ptolemaic system was still sanctioned by the Church. As a result, Galileo's enthusiasm soon brought him into trouble, and in 1616 he was expressly forbidden by Church authorities 'to hold or to defend' the heliocentric model. Thinking that he was still free to consider the notion - without specifically holding it or defending it - years later, in 1632, he published a new work, which had even been approved by the Church censors, entitled *Dialogue on the Great World Systems*, in which characters debated and discussed the Ptolemaic and Copernican systems.

The next year, at the age of 70, Galileo was brought before the Holy Office of the Inquisition. After a sham trial, he was threatened with torture and was forced to recant his work - to literally get down on his knees to formally renounce the condemned doctrine that the Earth revolves around the Sun, and to furthermore swear that he would turn in to the Inquisition any heretic who professed to believe in it. He probably remembered the fate of **Giordano Bruno** (1548 - 1600) - the Italian philosopher who, in 1600, was burned at the stake by the Inquisition for his belief that the universe was infinite and divine. Tragically, Galileo had no choice but to recant, and he was then forced to spend the rest of his life under house arrest.

I HATE IT WHEN I'M RIGHT!

Farther away from Rome, Galileo's contemporary, the German astronomer, **Johannes Kepler** (1571 - 1630), fared much better in his support of the Copernican system. Kepler worked as an assistant to the Danish astronomer, **Tycho Brahe** (1546 - 1601), at his pre-telescopic observatory in Prague, Czechoslovakia. By studying Brahe's precise planetary observations, Kepler was able to formulate his universal laws of planetary motion, which stated that the planets orbited the Sun in elliptical paths, and that their distance from the Sun determined the speed of their orbits.

One of the lingering problems with the Copernican system had been that it maintained that the planets orbited the Sun in perfectly circular orbits. In order to fit this belief with the observations, Copernicus himself had been compelled to employ the extraneous notion of epicycles - which had been one of the problems of the Ptolemaic system that he had originally tried to solve.

By discovering that planets actually have *elliptical* orbits and varying speeds, Kepler was finally able to do away with epicycles altogether, thus supporting the Copernican system - and solving its greatest problem at the same time!

One thing that continued to puzzle Kepler, however, was the question of why the planets orbited the Sun at all. By now, the notion of the heavenly, crystal-clear spheres seemed rather obsolete, to say the least. So what then held the universe together? Kepler had a theory that the Sun perhaps exerted some sort of magnetic force upon the planets to keep them in their orbits. This would explain why they moved faster when they were closer to the Sun, and slower when they were farther away from the Sun.

Kepler also held the belief that the Moon exerted some sort of force as well, which could be seen to affect the ocean's tides.

Inspired by Kepler's work, the English physicist, **Sir Isaac Newton** (1642 - 1727), took his ideas and ran with them. Newton determined that there was a force holding the universe together, but it wasn't magnetism - it was gravitation, the same force that makes an apple fall from a tree!

Newton's **Law of Universal Gravitation** stated that all objects exert a force of attraction upon all other objects, with *two* conditions: first, the larger the object, the greater the force of its attraction; second, the more distant the object, the weaker the force of its attraction.

Thus, the very force that grounds us to the Earth, keeping us from floating off into space, was the same force that pulled the Moon to the Earth, keeping it from flying off into space. Yet, clearly, gravitation wasn't the whole story. If it were, then the Moon should come crashing down to Earth.

To explain why it didn't, Newton appealed to the **Law of Inertia**. This law stated that all objects tend to remain in their current state - at rest or in motion - unless acted upon by another force. The reason why gravity doesn't pull the Moon down to Earth is because there is another force at work. The Moon's *momentum* - its prior state of motion - acts against the force of gravity, while gravity simultaneously acts against the force of its momentum. The two forces working together thus keep the Moon in its orbit. If one were to suddenly take away the Earth's gravitational pull, the Moon would go flying off into space in the direction of a straight line.

Newton then went on to say that the elliptical shapes of orbits are, in fact, the direct result of gravitation and inertia working together.

And just as the gravitational pull of the Earth, acting on the Moon's inertia, keeps the Moon in its elliptical orbit, so does the gravitational pull of the Sun, acting on the inertia of the planets, keep the planets in *their* elliptical orbits. Furthermore, because the force of an object's gravitational pull depends on its distance, planets closer to the Sun have faster orbits, and planets farther away from the Sun have slower orbits.

By providing an explanation for *how* and *why* the solar system works the way it does, Newton's work finally sealed the deal for the heliocentric system.

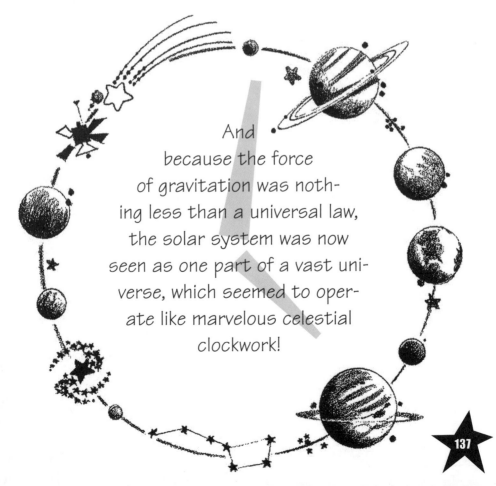

And because the force of gravitation was nothing less than a universal law, the solar system was now seen as one part of a vast universe, which seemed to operate like marvelous celestial clockwork!

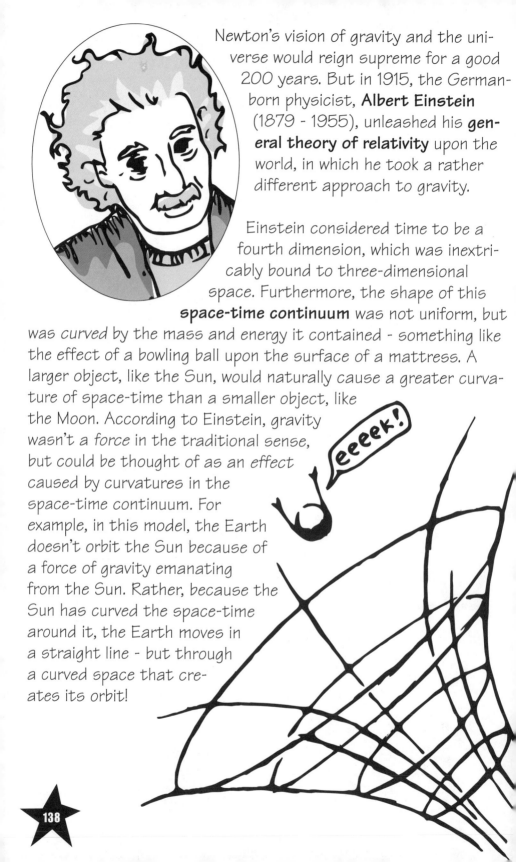

Newton's vision of gravity and the universe would reign supreme for a good 200 years. But in 1915, the German-born physicist, **Albert Einstein** (1879 - 1955), unleashed his **general theory of relativity** upon the world, in which he took a rather different approach to gravity.

Einstein considered time to be a fourth dimension, which was inextricably bound to three-dimensional space. Furthermore, the shape of this **space-time continuum** was not uniform, but was curved by the mass and energy it contained - something like the effect of a bowling ball upon the surface of a mattress. A larger object, like the Sun, would naturally cause a greater curvature of space-time than a smaller object, like the Moon. According to Einstein, gravity wasn't a *force* in the traditional sense, but could be thought of as an *effect* caused by curvatures in the space-time continuum. For example, in this model, the Earth doesn't orbit the Sun because of a force of gravity emanating from the Sun. Rather, because the Sun has curved the space-time around it, the Earth moves in a straight line - but through a curved space that creates its orbit!

eeeek!

In most cases, the predictions of classical physics according to Newton will be identical to the predictions of the theory of relativity according to Einstein - but not always. As one example, Einstein predicted that the light from distant stars travels in a straight line, but can also be bent by curvatures in space-time, such as the curvature caused by the Sun. Of course, normally one can't see the light from a star in the vicinity of the Sun's brightness. However, in an experiment to test this theory, photographs were taken of the Sun during a total solar eclipse, in which nearby stars could be seen. These were then compared to photographs taken of the same stars in the night sky, far away from the Sun. The results not only confirmed that the Sun's gravitational effect changed the positions of the light from the stars - but that it did so to practically the exact degree that Einstein had predicted.

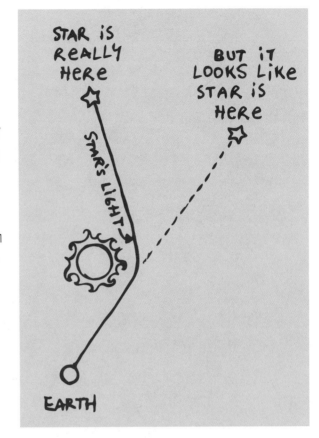

STAR IS REALLY HERE

BUT iT LOOKS LiKE STAR iS HERE

STAR'S LIGHT

EARTH

ABSOLUTELY AMAZING!!!

iSAAC NEWTON

At the time, Einstein, like most people, continued to assume that the universe still operated with the same clockwork-like precision that Newton had envisioned. In other words, the universe was still considered to be a fairly mechanical place. One thing that Einstein had *failed* to predict, however, was a consequence of his theory that was later brought to his attention. Namely, if the shape of the entire space-time continuum was curved, then the gravitational effects would have to imply that the universe was either contracting into a *big crunch*, or expanding away in all directions - but it couldn't be standing still. Initially, Einstein was sure that this was a mistake, but, in fact, he was wrong!

ME... WRONG???

At just about the same time, during the mid-1920's, the American astronomer, **Edwin Hubble** (1889 - 1953), was working at the Mt. Wilson Observatory in Pasadena, California. Completely unaware of Einstein's general theory, Hubble hit upon the greatest discovery of 20th century science: the expansion of the universe. The details of this discovery once again changed the way we looked at the cosmos, indicating that the universe must have had a beginning at a singularity in space-time - what we now know of as the

BIG BANG.

Hubble observed that the space between clusters of galaxies was expanding, but that galaxies and galaxy clusters were not expanding in the same way. This is because galaxies and galaxy clusters are held together by gravitation - which would seem to be stronger than the force of universal expansion. The curious thing, however, was that it appeared that all distant galaxies were expanding away from our own galaxy in every direction.

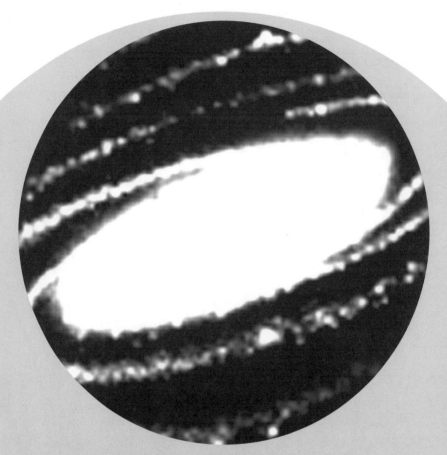

COULD THIS HAVE POSSIBLY
MEANT THAT OUR GALAXY WAS,
IN FACT, THE VERY CENTER OF
THE UNIVERSE AFTER ALL?

The astonishing answer - as it turns out - is that *there is no center of the universe!*

he initial singularity that gave birth to our universe contained all
natter, energy, time, and space, and nothing else existed outside
f it. It would thus be incorrect to say that the universe originat-
d from a single point in space, because the universe *contains* all
he space there is - and always has! As a result, almost all parts
f the universe are simultaneously expanding away from all other
arts of the universe.

his somewhat difficult concept can be better understood if we
ompare the shape of the universe to the surface of a balloon. As
balloon is inflated, from any given point on the surface of the
alloon, all other points will expand away in every direction. And
ust as every point on the balloon can be thought of as the cen-
er of the balloon, so can every point in the universe be thought
f as the center of the universe!

And
in a sense, this dis-
covery brings our expedition
right back to where we started.

For now there really isn't any logical
reason why we *can't* go back to thinking
of ourselves as the center of the
universe:

The only condition is that we just
have to be willing to share this
honor with the rest of the
cosmos!

Like life itself, science is a journey, not a destination. As social anthropologist Sir James George Frazer put it in *The Golden Bough* (1915): "The advance of knowledge is an infinite progression towards a goal that forever recedes." As such, our intellectual aim should never be to *complete* our understanding, but merely to continue to *increase* our understanding. This brief portrait of our current stage of understanding, as presented here, is but a snapshot in time. Our journey continues, and may we never stop exploring.

[*The Illustrated Golden Bough, A Study in Magic and Religion*, The Softback Preview, UK, 1996.]

THE END

Acknowledgements

First and foremost, I'd like to thank the nine muses for the inspiration, especially Calliope and Urania. I'd also like to acknowledge my appreciation and love for my parents. Heaps of gratitude to my sister, Sarah, for her stellar illustrations. Very special thanks go to Dr. Jim Zimbelman, at the Smithsonian's National Museum of Air and Space, for his invaluable feedback and encouragement. And finally, lots of love to Laurie, without whom this could not have been possible.

Recommended for Further Reading

Barrow, John D. *The Origin of the Universe*. Basic Books, 1994.

Davies, Paul. *The Fifth Miracle: The Search for the Origin and Meaning of Life*. Simon & Schuster, 1999.

Ferris, Timothy. *Coming of Age in the Milky Way*. Doubleday & Company, Inc., 1989.

Ferris, Timothy. *The Whole Shebang: A State of the Universe(s) Report*. Touchstone, 1997.

Harrington, Philip S. *Eclipse! The What, When, Why & How Guide to Watching Solar and Lunar Eclipses*. John Wiley & Sons, Inc., 1997.

Hawking, Stephen. *A Brief History of Time: From the Big Bang to Black Holes*. Bantam Books, 1988.

Moore, Patrick. *Atlas of the Universe*. Cambridge University Press. 1998.

O'Byrne, John (editor). *Advanced Skywatching*. The Nature Company Guides, Time Life Inc. 1997.

Sagan, Carl. *Cosmos*. Ballantine Books, 1980.

Sagan, Carl, and Ann Druyan. *Comet*. Random House, 1985.

Trefil, James. *Reading the Mind of God: In Search of the Principle of Universality*. Charles Scribner's Sons, 1989.

Index

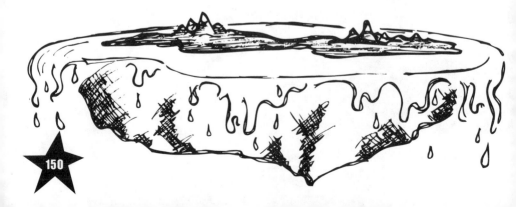